Green Computing Technologies and Computing Industry in 2021

Edited by Albert Sabban

Published in London, United Kingdom

IntechOpen

Supporting open minds since 2005

Green Computing Technologies and Computing Industry in 2021
http://dx.doi.org/10.5772/intechopen.92470
Edited by Albert Sabban

Contributors
Satish Shamrao Kola, Mohammad Idrees, Naqui Siddiqui, Roshan Nasare, Chandrashekhar Gorakhnath Devkate, Wilson Nwankwo, Paschal Uchenna Chinedu, Zebun Nasreen Ahmed, Mohammad Tanvir Kawser, L.S. Supian, Dr. Mohd Syuhaimi Ab-Rahman, Norhana Arsad, Hadi Guna, Khadijah Ismail, Nani Fadzlina Naim, Harry Ramza, Nik Ghazali Nik Daud, Fortune Ganda, Rufaro Garidzirai, Albert Sabban

First published in London, United Kingdom, 2021 by IntechOpen
IntechOpen is the global imprint of INTECHOPEN LIMITED, registered in England and Wales, registration number: 11086078, 5 Princes Gate Court, London, SW7 2QJ, United Kingdom
Printed in Croatia

British Library Cataloguing-in-Publication Data
A catalogue record for this book is available from the British Library

Additional hard and PDF copies can be obtained from orders@intechopen.com

Green Computing Technologies and Computing Industry in 2021
Edited by Albert Sabban
p. cm.
Print ISBN 978-1-83968-752-5
Online ISBN 978-1-83968-753-2
eBook (PDF) ISBN 978-1-83968-754-9

For EU product safety concerns:
IN TECH d.o.o., Prolaz Marije Krucifikse Kozulić 3, 51000 Rijeka, Croatia, info@intechopen.com or visit our website at intechopen.com.

Meet the editor

Dr. Albert Sabban has a Ph.D. in Electrical Engineering from the Faculty of Electrical and Computer Engineering, University of Colorado Boulder (1991), and an MBA from the Faculty of Management, Haifa University, Israel (2005). He holds both a BSc and MSc, Magna Cum Laude, from the Electrical and Computer Engineering Faculty at Tel Aviv University. He is a senior lecturer and researcher in electrical and computer engineering at several colleges. From 1976 to 2008 he worked at RAFAEL, an Israeli defense technology company, as a senior researcher, group leader, and project leader. From 2007 to 2021 he worked as a radio frequency (RF) specialist project leader at high-tech companies and a senior lecturer and researcher. He has published more than 100 research papers and holds a number of US patents. Dr. Sabban has written and edited nine books on compact wearable systems and green technologies. He has also written books on electromagnetics and wide-band microwave technologies and six book chapters on wearable printed systems and green technologies.

Contents

Preface

The book is divided into the following four sections:

1. "Introduction"
2. "Green Programmable and Computing Systems"
3. "Green Technologies and Devices"
4. "Green Computing Industry and Applications"

The continuous growth and development of cellular wireless communication systems over the last thirty years has resulted in most of the world's population owning smartphones, smartwatches, iPads, and other devices. Almost every person has more than one computing device, thus there are billions of devices consuming megawatts of power. Therefore it is crucial to design energy-efficient, or "green," technologies and devices. Moreover, due to the rapid advance of technology, many people upgrade their devices frequently, which leads to a great number of discarded and unwanted electronics. Computers and electronic waste are filling up landfills at an alarming rate. These electric devices contain hazardous and toxic materials that pollute the environment and endanger the health of local communities. Green computing and electronic technologies are employed to combat this environmental pollution. Recycling electronic waste, old batteries, plastics, and so on help to decrease the environmental pollution. Renewable energy is also a major factor in decreasing the environmental pollution. In the last century, the world has suffered from rapid and negative environmental changes, including severe droughts, seawater acidification, rising seawater levels, increased depletion of groundwater reserves, and the global rise of earth temperature. The rapid spread of diseases, viruses, macro-parasites, and the extinction of animal species are the direct result of environmental changes. Most of these changes are irreversible. It is obvious in 2021 to most of the world that green computing technologies are crucial in protecting our universe from environmental hazards and pollution. Countries, governments, communities, and citizens should act rapidly to save the planet.

Green computing is the environmentally responsible use of computers and their resources. Green computing may be considered as the research and study of developing, designing, engineering, producing, using, and disposing of computing modules and devices to reduce environmental hazards and pollution. Computer designers, developers, manufacturing companies, and vendors are investing in developing green computing modules and devices by reducing the use of hazardous materials and improving the recycling process of computing and digital modules. Green computing is also known as green information technology (green IT).

Green computing main topics and initiatives include:

- Energy consumption: Minimizing the electricity consumption of computers and their peripheral devices and using them in an eco-friendly manner.

- Green disposal: Disposing and recycling unwanted computing devices.
- Green development and design: Designing and developing energy-efficient computers, servers, printers, projectors, and other digital devices, and using energy harvesting technologies in the computer industry.
- Green manufacturing: Recycling electronic components and modules during the manufacturing of computers and their peripheral devices. Minimizing waste during the manufacturing of computers and other subsystems to reduce the environmental impact of these activities.

This book examines innovation in green computing technologies and industry. Each chapter covers significant information to enable engineers, students, and scientists from all areas to follow and understand the topics presented in the book. The book presents new subjects and innovations in green computing technologies and in green computing and electronics industries. The main topics presented in the book include:

- Chapter 1 "Introductory Chapter: Green Computing Technologies and Industry in 2021"
- Chapter 2 "Wideband Systems with Energy Harvesting Units for 5G, Medical and Computer Industry"
- Chapter 3 "Internet Connectivity in Building Interiors: Architecture and Sustainability Considerations"
- Chapter 4 "Polymer Optical Fiber Splitter Using Tapered Techniques for Green Technology"
- Chapter 5 "Eco-Friendly, Green Approach for Synthesis of Bio-Active Novel 3-Aminoindazole Derivatives"
- Chapter 6 "Green Computing: A Machinery for Sustainable Development in the Post-Covid Era"
- Chapter 7 "The Environmental Influence of Tax Regimes in Selected European Union Economies"

I want to thank all the chapter authors for their excellent contributions.

Albert Sabban
Kinneret Academic College,
Galilee, Israel

Ort Braude College,
Karmiel, Israel

Section 1

Introduction

Chapter 1

Introductory Chapter: Green Computing Technologies and Industry in 2021

Albert Sabban

1. Introduction to green computing technologies

Almost every person has more than one computing device. So, there are in our world milliards of computing devices that consume a huge amount, of megawatts. It is crucial to design energy-efficient computers, servers and Tablets. Computer engineers should design and manufacture green computers, servers, Tablets, and other computing devices. The computer industry design and manufacture every year new computing devices. Appropriately disposing, recycling of unwanted computing devices and electronic equipment is a crucial task in creating a green environment.

Green computing is the environmentally responsible use of computers and their resources. Green computing may be considered as the research and study of developing, designing, engineering, producing, using, and disposing of computing modules and devices to reduces environmental hazards and pollution. Computer designers, developers, manufacturing companies and vendors are investing in developing green computing modules and devices by reducing the use of hazard materials and improving the recycling process of computing and digital modules. Green computing is also known as green information technology (green IT).

In 1992 the Environmental Protection Agency (EPA) launched the Energy Star program that boosted Green computing practices.

Green computing practices include the development of environmentally green production practices, efficient energy consumption computers, usage of energy harvesting technologies, and enhancing disposal and recycling procedures.

Green computing main topics and initiatives:

- **Energy Consumption** - Minimizing the electricity consumption of computers and their peripheral devices and using them in an eco-friendly manner

- **Green disposal:** Disposing and recycling, unwanted computing devices

- **Green development and design:** Design and development of energy-efficient computers, servers, printers, projectors, and other digital devices. Usage of energy harvesting technologies in the computer industry.

- **Green manufacturing:** Recycling electronic components and modules during the manufacturing of computers and their peripheral devices. Minimizing waste during the manufacturing of computers and other subsystems to reduce the environmental impact of these activities

IntechOpen

The book presents new subjects and innovation in green computing technologies and in green computing and electronics industries. The main topics presented in the book are listed.

- Green computers, electronics, and computing technologies
- Green renewable energy in computers and electronics
- Developing and manufacturing of energy-efficient computers, servers and Tablets
- Recycling and green disposal of computers and digital devices
- Low electricity consumption computers, smartphones, servers and Tablets
- Green production of green computers and digital devices

It is obvious in 2021 to most of the world population that Green Computing Technologies are crucial in protecting our universe from environmental hazards and pollution. In the last century, the world has suffered from rapid changes in climate, water pollution, and air pollution. The universe suffers from severe droughts, seawater acidification, rising seawater levels, increased depletion of groundwater reserves, air pollution, rivers pollution, and global rise of earth temperature. The rapid spread of diseases, viruses, macro-parasites, and the extinction of animal species are the direct result of environmental changes. Most of these changes are irreversible. Countries, governments, communities, and citizens should act rapidly to save the planet. In the last decade we notice that kids and young people lead around the world initiatives to decrease air pollution, water pollution and climate changes.

The private cars industry, electronic industry, high tech computing industry, and the communication industry in the last century depleted and ruined the world natural resources. Computers waste, electronic waste, plastics, and food garbage contain dangerous chemicals that pollute air, soil, rivers water, sea water, and groundwater. These toxic materials cannot be removed from fishes, vegetables, fruits, and other food products. These toxic chemicals can be found in fruits, vegetables and other food crops, fishes, meat, and corps grown on polluted soil. Toxic waste, polluted water, polluted air, and climate changes affect severely first children health and grownup health. The universe oceans, rivers, seas, and lakes suffer from plastic and chemical waste. Chemical toxics and plastic waste kill the ocean and sea habitats in our planet. Hazard materials, chemical toxics, and plastic waste kill fishes, birds, and other creatures. Plastic waste poses choking and strangulation hazards. Plastic particles are making their way into the bottom of the food chain. Fishes and other species swallow plastic waste and become contaminated. These creatures are subsequently harvested for human consumption. The contaminated fishes and creatures, hazard toxics, are served in our plate and reach our stomach.

Green computing, green electronics, green energy, recycling, and waste management and are the important crucial topics and challenges in green technologies, green industry, green research and in green innovation. The book is divided to three sections. Section one presents innovations in green computing and electronic systems. Section two presents green devices and technologies. Innovation in green computing technologies and computing industry are presented in section 3.

2. Green computing and electronics technologies

Green electronics and computing are in continuous growth in the last twenty years. The communication, wireless communication, electronics, and computing industries are facing increasing pressure from governments and legislation to remove toxic and hazardous material from their devices. There is a continual worldwide environmental movement to use green materials, green energy, and green components in production of communication, electronic devices, and computers. There are global series of activities that outline targets for electronic devices re-use and recycling. In green computing and green electronic industries, the use of hazardous materials such as cooper, lead, plastic materials, and other toxic materials is not permitted or limited to decrease pollution and to improve recycling process.

Computers, communication devices, and electronics manufacturers are encouraged and challenged to develop electronic devices that are green and environmentally friendly. Green communication and electronics manufacturing focuses on using green materials, renewable energy, using efficient green devices, using energy harvesting units, reduction number of parts in modules and systems, recycling green materials, recycling energy, and recycling mechanical and electronic components. The book will describe and present development and production of green efficient devices that use green energy in computing and electronic industry. In this regard, energy harvesting technologies that can be employed to produce, recycle, and store green energy are presented in this book. On the other hand, it will present green computing systems and devices that are important factors in developing greener computing and electronic industry. Computers, cell phones and other electronic devices contain toxic hazardous materials that endanger the environment and, consequently, threaten the human health. Recycling of computer and electronic waste minimize the amount of toxic hazardous materials in the environment. Computers developers and manufacturers should handle and take care of electronic waste seriously with first majority. Computers, smart phones, and electronics manufacturers are challenged to reduce the amount, of computers and electronic waste and the number of parts in the system, by designing products that minimize the amount, of parts in the system and of harmful substances and to use parts and components that can be recycled. Computers developers and manufacturers are encouraged to develop products with a longer lifecycle. This book intends to provide the reader with a comprehensive presentation of green computing, green environmental technologies as a global universal standard for projects managers, developers, system engineers and manufacturers.

Main Activities in Green Electronics and Computing

- Green energy
- Minimization in energy consumption
- Green materials
- Electronic waste
- Recycling

3. Green computing- cloud storage and computing services

Cloud storage cut energy consumption, computing expenses such us storage maintenance tasks, and purchasing additional storage capacity. Cloud storage provides greener computing services. Cloud storage is a service package in which data is stored, managed, backed up remotely and made available to users over a network and internet services. Cloud storage is based on a virtualized infrastructure with accessible interfaces. Cloud-based data is stored in servers located in data centers managed by a cloud provider. A file and its associated metadata are stored in the server by using an object storage protocol. The server assigns an identification number, ID, to each stored file. When file needs to be retrieved, the user presents the ID to the system and the content is assembled with all its metadata, authentication, and security. The most common use of cloud services is cloud backup, disaster recovery, and archiving infrequently accessed data. Cloud storage providers are responsible for keeping the data available and accessible, and the physical environment protected and running. People and organizations buy or lease storage capacity from the providers to store and archive data files. Cloud storage services may be accessed via cloud computers and web services that use application programming interface, API, such as cloud desktop storage and cloud storage gateways.

There are three main cloud-based storage architecture models public, private and hybrid.

Public cloud storage services provide a multi customer storage environment that is most suited for data storage. Data is stored in global data centers with storage data spread across multiple regions or continents.

Private cloud storage provides local storage services to a dedicated environment protected behind an organization's firewall. Private clouds are appropriate for users who need customization and more control over their data.

Hybrid cloud storage is a mix of private cloud and third-party public cloud services with synchronization between the platforms. The model offers businesses flexibility, and more data deployment options. An organization might, for example, store actively used and structured data in a local cloud, and unstructured and archival data in a public cloud. In recent years, a greater number of customers have adopted the hybrid cloud storage model. However, hybrid cloud storage presents technical, business and management challenges. For example, private workloads must access and communicate with public cloud storage suppliers, so solid network connectivity and compatibility are very important issues in this case.

Advantages of Cloud Storage

- By using cloud storage companies can cut their energy consumption, computing expenses such us storage maintenance tasks, and purchasing additional storage capacity.

- Cloud storage provides users with immediate access to a broad range of resources and applications hosted in the infrastructure of another web service interface.

- Cloud storage can provide the benefits of greater accessibility, rapid deployment, strong protection for data backup, reliability, archival and disaster recovery purposes.

- Cloud storage is used as a natural disaster proof backup. Usually there are at least two backup servers located in different places around the world.

- Cloud storage can be used for copying virtual machine images from the cloud to a desired location or to import a virtual machine image from any designated location to the cloud image library.

- Storage availability and data protection are provided by cloud storage services. So, depending on the application, the additional technology efforts, and cost to ensure availability and protection of data storage can be eliminated.

Disadvantages of Cloud Storage

- Decrease in the security level of the stored data.

- Increasing the risk of unauthorized physical access to the data.

- In cloud-based architecture, data is replicated and moved frequently so the risk of unauthorized data recovery increases dramatically.

- It increases the number of networks that the data should travel over them. This disadvantage does not exist in a local area network (LAN) or in a storage area network.

- Data stored on a cloud requires a wide area network.

- A cloud storage company have many customers and thousands of servers. Therefore, a larger team of technical staff with physical and electronic access to almost all, the data at the entire facility. Encryption keys that are kept by the service user, as opposed to the service provider, limit the access to data by service provider employees. An amount, of keys have to be distributed to users via secure channels for decryption. The keys should be securely stored and managed by the users in their devices. Storing these keys requires expensive secure storage.

- Cloud storage is a rich resource for both hackers and national security agencies. The cloud store data from many different users and organizations. Hackers see it as a very valuable target.

- Cloud storage sites have faced lawsuit from the owners of the intellectual property uploaded and shared in the site. Piracy and copyright problems may be enabled by sites that permit file sharing.

- Cloud storage companies are not permanent and the services and products they provide can change.

- Cloud storage companies can be purchased by other foreign larger companies, can go bankrupt and suffer from an irrecoverable disaster.

Cloud Computing.

Cloud computing is an internet computing service that provides shared computer resources, stored data to computers and other devices, and computer processing resources on demand. **Cloud computing** provide access to a shared pool of configurable computing devices such as servers, computer networks, data storage devices, computing applications, and other services. Cloud computing relies on sharing of computing resources. **Cloud computing** services can be rapidly provisioned and released with minimal management effort. Cloud computing and storage solutions provide users and enterprises with various capabilities to store and process their data in privately owned data centers. Cloud computing allows companies to avoid high infrastructure costs such as servers and expensive software. Cloud computing allows companies to get their applications up and running faster, with improved manageability and less maintenance costs. Information technology teams can rapidly adjust resources to meet unpredictable business demands. Cloud computing applies high-performance computing power to perform tens of trillions of computations per second. Cloud computing allow organizations to focus on their core businesses instead of spending time and money on computer networks. **Cloud computing** and storage solutions cut companies energy consumption.

4. Green renewable energy

Solar energy, wind energy, water energy, and biology fuel are examples of green renewable energy. Nuclear energy, fuel, hydrogen, coal, natural gas, and oil are nonrenewable energy resources. Light, wind, electromagnetic energy, are employed to produce green renewable energy.

4.1 Solar energy

The radiation of the sun is used to produce solar energy. The sun is an infinite limitless source of solar energy. Solar cells convert energy from natural light into electrical energy through the process of photovoltaics. Using solar energy to generate electricity minimize the consumption of coal and fuel. Using solar energy results in a significand reduction in air and environmental pollution. Solar energy farms are presented in **Figure 1**.

Advantage of Green Solar Energy

- does not emit green-house gases
- Solar energy is Green and clean energy.
- Solar energy offers decentralization of power.
- Solar energy is environment friendly. Solar energy does not rely on constantly mining raw materials. Solar energy does not result in the destruction of forests and eco-systems that occurs with many fossil fuel operations.

Figure 1.
Solar energy farms.

- Solar panels produced today carry a 25 to 30 years warranty.
- Solar energy is not degradable.
- Solar light is a free natural resource
- Solar energy does not require expensive and ongoing raw materials like oil, coal, or gas. Solar energy requires significantly lower operational labor than conventional power production. Raw materials should not be constantly extracted, refined, and transported to the power plant.
- Oil, coal, and gas used to produce conventional electricity is often transported cross-country or internationally. This transportation has additional costs, including monetary costs, and pollution costs of transport. These costs are avoided with solar energy.
- On grid means If the house remains connected to the state electricity facility it is called on grid connection. If the house has no connection to the state electricity facility it is called off grid connection. In this case the facility, house, company, business, or a community is relying only on solar energy. The ability to produce electricity off the grid is an important advantage of solar energy for isolated communities, facilities, and rural areas. Solar energy can be produced on or off the grid. Installing power long power lines are significantly difficult and very expensive in these rural areas.
- The sun is an unlimited free commodity that can be sourced from several locations. One of the major advantages of solar energy is the ability to bypass the linkage between politics status and energy price. However, in the fuel markets there is a strong linkage between politics status and energy price. Price of Solar energy is less affected to price manipulations of politics status, war, and international relations. However, politics status manipulations doubled the price of fuel in the past fifty years.

Disadvantage of Solar Energy

- Depend on the weather and sun light
- Expensive installation and maintenance
- Consume large area
- Development and production of solar cite is expensive
- solar energy production is relatively inefficient. The efficiency of solar panels is less than 30%. This means that a huge amount of surface area is required to produce adequate electricity.
- Solar electricity storage technology is expensive, bulky, and more appropriate to small scale home solar panels than large solar farms.

4.2 Wind energy

Global efforts to combat climate change, such as the Paris Agreement, renewable energy is seeing a continues growth. Cumulative wind capacity around the world increased from around 20,000 megawatts to more than 500,000 megawatts in the last fifteen years. Wind power is a popular sustainable, renewable source of power that has a small impact on the environment compared to burning coal or fuel. Wind energy uses the wind to provide mechanical power through wind turbines to turn electric generators for electrical power. The wind kinetic energy is used to operate electric turbines and windmills. The turbine's blades spin clockwise when the wind blows, capturing energy. When the wind blows the main shaft of the wind turbine, connected to a gearbox within the nacelle, is triggered to spin. The gearbox sends the wind energy by the gearbox to the generator. The wind energy is converted to electricity. However, windmills cannot be operated in a residential area. Offshore wind turbines provide steady, reliable clean energy in several countries. **Figure 2** presents wind energy cites. Wind is a clean source of renewable energy that produces no air or water pollution. Wind energy disadvantages include complaints from locals that wind turbines are ugly and noisy. The turbines rotating blades kill birds and bats.

Advantage of Wind Energy

- Wind energy is a green and clean energy. Wind energy turbines operation does not directly emit any CO2 or greenhouse gases
- Wind energy is plentiful, readily available, and capturing its power does not deplete our valuable natural resources.
- Not degradable
- Wind power is cost-effective in many regions. Wind energy is cheap.
- Wind energy offers decentralization of power.
- Wind energy is environment friendly. Wind energy does not rely on constantly mining raw materials. Wind energy does not result in the destruction of forests and eco-systems that occurs with many fossil fuel operations.
- Wind is a free natural resource
- Wind energy does not require expensive and ongoing raw materials like oil, coal, or gas. Wind energy requires significantly lower operational labor than conventional power production. Raw materials should not be constantly extracted, refined, and transported to the power plant.

(a) (b)

Figure 2.
a. Country wind energy farm b. offshore wind energy cite

- The wind is an unlimited free commodity that can be sourced from several locations. One of the major advantages of wind energy is the ability to bypass the linkage between politics status and energy price. However, in the fuel markets there is a strong linkage between politics status and energy price. Price of wind energy is less affected to price manipulations of politics status, war, and international relations. However, politics status manipulations doubled the price of fuel in the past fifty years.

- Oil, coal, and gas used to produce conventional electricity is often transported cross-country or internationally. This transportation has additional costs, including monetary costs, and pollution costs of transport. These costs are avoided with solar energy.

Disadvantage of Wind Energy

- Depend on the weather and wind velocity

- Cannot be operated in a residential area

- Consume large area

- Development and production of wind energy cite is expensive, land is expensive.

- Noisy

- Dangerous to birds, bats, and animals

Table 1 presents global wind farms database.

4.3 Hydropower, water energy

Waterfalls and fast running water flow may be used to generate electric energy. Water Energy cite is presented in **Figure 3**. Water flow kinetic energy is converted to electric energy. In the late 19th century, hydropower became a source for generating electricity. The first commercial hydroelectric power plant was built at Niagara Falls in 1879. In 1881, streetlamps in the city of Niagara Falls were powered by hydropower. Hydroelectricity can be used to store energy in the form of potential energy between two areas with different heights with pumped-storage hydroelectricity. Water is pumped uphill into cites during periods of low demand. This energy can be released to generate energy when demand is high.

Hydropower Energy Sources

- Rain fall generates water flow

- Waterfalls generate hydropower energy

- High water levels, low water levels and water stream can be used to generate hydropower energy

Hydropower Energy Advantages

- Hydroelectricity is green and clean energy compared to other sources of energy.

- Hydroelectricity is not degradable
- Water is a free natural resource
- Hydroelectricity is cheap energy compared to other sources of energy
- Hydroelectricity does not pollute the environment

Disadvantage of Hydropower Energy

- Hydropower energy cites depend on rain fall, snow, the weather, and water stream

Area	Listed capacity	Entries	Prices
World	884.2 GW	32,293	1250 €
Offshore	344.5 GW	986	200 €
Africa	8.8 GW	125	75 €
Americas	207.1 GW	3,165	400 €
Asia	328.2 GW	4,612	400 €
Europe	324.4 GW	24,195	800 €
Oceania	15.7 GW	196	100 €
Germany	70,800 MW	11,131	350 €
Brazil	36,476 MW	678	205 €
USA	136,490 MW	1,778	350 €
China	205,567 MW	2,528	350 €
United Kingdom	55,171 MW	1,171	345 €
France	22,252 MW	1,574	350 €
India	33,399 MW	914	205 €
Japan	25,107	391	105 €

Table 1.
Global wind farms database.

Figure 3.
Hydropower energy cite.

- Difficult to be installed and operated in a residential area
- Hydropower energy cites usually are limited to areas with waterfalls
- Hydropower energy cites consume large area
- Development and production of water hydroelectricity is expensive.

4.4 Energy harvesting

The continuous growth in production of portable RF systems increase the consumption of batteries and electrical energy. Batteries and conventional electrical energy increase the environmental pollution. In last twenty years, the trend of using free space energy such us light, heat, electromagnetic energy, vibration, muscle motion and other energy sources, has become very attractive and useful. Several inventions and methods to produce electricity from green energy sources have been presented, see [1–3]. RF harvesting may be useful and recharge batteries only if we collect as much RF energy as possible. In this case, energy harvesting units can eliminate the need to charge batteries by using electrical cables. It is crucial to harvest RF energy from several RF devices and systems. Wideband or multiband antennas should be used to harvest as much electromagnetic energy as possible. The energy harvesting antenna must meet the system requirements. Due to low electromagnetic energy densities in free space, efficient antennas should be used. The antennas should radiate efficiently at a specific frequency range and polarization. The antenna should receive efficiently electromagnetic waves from a wide angular angle. To meet this requirement the antenna radiation pattern should have a wide beam width. Printed antennas were used to harvest RF energy as presented in the literature, [4–10]. Electromagnetic energy harvesting systems capture waves propagating in the air. This RF energy is stored and used to recharge batteries and other electrical devices. In the last decade there is a huge increase in the amount of RF power in the air. The amount, of electromagnetic waves in the air in 2017 was 11 Exa-bytes per month. The amount, of electromagnetic waves in the air in 2019 was 33 Exa-bytes per month. However, the predicted amount, of electromagnetic waves in the air in 2025 will be around 165 Exa-bytes per month, **Table 2**. **Table 2** presents the expected amount of RF wave in the air for 2G, 3G, 4G, and 5G networks. 5G devices are forecast to account for 45% of the universe mobile data traffic by 2025.

Energy harvesting system is shown in **Figure 4**. The electromagnetic harvesting unit consists of an antenna, a rectifying circuit, and a rechargeable battery. The RF harvesting energy system operates as a Dual Mode harvesting unit. The Low Noise

Year	Total amount of RF wave in free space EB per month	Amount of RFwave in free space EB per month 2G, 3G, 4G	Amount of RF wave in free space EB per month 5G
2017	11	11	0
2019	33	33	0
2021	60	50	10
2023	100	75	25
2025	164	85	79

Table 2.
Amount, of electromagnetic waves in the air from 2017 up to 2025.

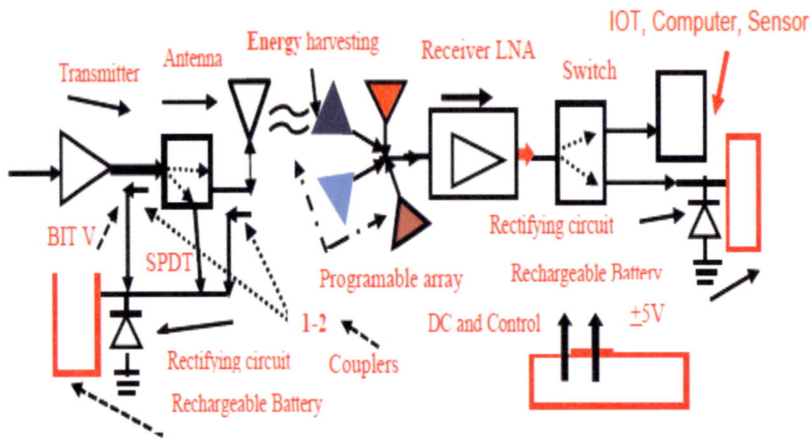

Figure 4.
Dual mode energy harvesting concept.

Amplifier, LNA, is part of the RF system. The LNA DC bias voltages are supplied by the DC unit of the RF system. The programmable array consists of 3 to 4 antennas that can harvest RF energy from around 140 MHz up to 18GHz. The total received electromagnetic energy is transformed to DC power. The harvesting system consists of antennas, matching and feed networks, rectifying circuit, and a rechargeable battery. The RF harvesting system function as a dual mode harvesting unit. The harvesting unit can be part of a computing network, server, communication system, medical, IOT, computer, and smartphone. The LNA bias voltages are supplied by the RF system. The energy coupled to the built in test port, -20 dB, may be used to recharge a battery.

5. Recycling

Recycling is the process of collecting and processing used materials that would otherwise be thrown away as trash. Recycling conserves raw materials and saves the additional energy that manufacturers would use producing new products from scratch. The continuous growth in development of cellular wireless communication systems over the last thirty years has resulted that most of the world population owning laptops, computers, smartphones, smart watches, I-pads, and other devices. As a result, the number of unwanted electronic devices is huge. With this huge number of devices being produced and discarded, a new environmental disaster strikes our planet. Electronic waste or discarded old electronics are filling up landfills and trash storage areas. These electronics waste contain hazardous materials and toxic materials that endanger the environment and the health of local communities. This electronics waste increase air and water pollution. Green electronics and computing technologies are used to decrease environmental pollution. Recycling of computers waste, electronic devices waste, old batteries, plastic waste, and bottles decrease environmental pollution. Recycling saves original raw materials. Recycling reduce waste and reduce pollution. Computers recycling waste is shown in **Figure 5**.

Benefits of Recycling

- Conserves natural resources such as cooper, steel. Water and other minerals
- Minimize energy consumption by using recycled devices and parts

Figure 5.
Computers waste recycling cite.

- Increases economic security by tapping a domestic source of materials
- Prevents pollution by reducing the need to collect new raw materials
- Reduces the amount of waste sent to landfills
- Saves energy needed to produce new components and materials
- Supports manufacturing companies and conserves valuable resources
- Create jobs in the recycling and manufacturing industries in the United States
- Price of recycled devices is cheaper

6. Innovations and challenges in green computing technologies

In the last fifty years, the world has suffered from significantly rapid changes in the planet weather, water pollution, and air pollution. The private cars industry, electronic industry, high tech computing industry, and the communication industry in the last century depleted and ruined the world natural resources. The universe suffers from severe droughts, seawater acidification, rising seawater levels, increased depletion of groundwater reserves, air pollution, rivers pollution, and global rise of earth temperature. The rapid spread of diseases, viruses, and the extinction of animal species are the result of environmental changes. Most of these changes are irreversible. Computers waste, electronic waste, plastics, and food garbage contain dangerous chemicals that pollute air, soil, rivers water, sea water, and groundwater. These toxic materials cannot be removed from fishes, vegetables, fruits, and other food products. These toxic chemicals can be found in fruits, vegetables and other food crops, fishes, meat, and corps grown on polluted soil. Toxic waste, polluted water, polluted air, and climate changes affect severely first children health and grownup health. The universe oceans, rivers, seas, and lakes suffer from plastic and chemical waste. Chemical toxics and plastic waste kill the ocean and sea habitats in our planet. Hazard materials, chemical toxics, and plastic waste kill fishes, birds, and other creatures. We should encourage countries, governments, communities, and citizens to act rapidly to save the universe. We should encourage companies to use green materials, green energy, and green components in production of communication, electronic devices, and computers. We should encourage global activities to re-use and recycle computers and electronic waste. In green computing and green electronic industries, the use of hazardous materials such as cooper, lead, plastic materials, and other toxic materials should be limited to decrease pollution.

Computers, communication devices, and electronics manufacturers should be encouraged to develop electronic devices that are green and environmentally friendly. Computing and electronics manufacturers should use green materials, renewable energy, efficient green devices, and energy harvesting systems.

Future Computing and Electronics Green Technologies Challenges and Innovations

- Using renewable and harvesting energy in computing and electronic devices
- Using recycled components and devices in manufacturing computers and electronic devices
- Design and development of green computers and electronic devices
- Producing cheap solar green energy
- Recycling most of the computers and electronics waste
- Development and production of green cars and airplanes
- We should encourage peoples, young people, and kids to lead around the world initiatives to decrease air pollution, water pollution and climate changes.

Author details

Albert Sabban[1,2]

1 Kinneret Academic College, Galilee, Israel

2 Ort Braude College, Karmiel, Israel

*Address all correspondence to: avi_111@netvision.net.il

References

[1] J. A. Paradiso and T. Starner, "Energy scavenging for mobile and wireless electronics," IEEE Pervasive Computing, vol. 4, no. 1, pp. 18-27, 2005.

[2] C. Valenta and G. D. Durgin, "Harvesting wireless power: survey of energy-harvester conversion efficiency in far-field, wireless power transfer systems," IEEE Microwave Magazine, vol. 15, no. 4, pp. 108-120, 2014.

[3] P. Nintanavongsa, U. Muncuk, D. R. Lewis, and K. R. Chowdhury, "Design optimization and implementation for RF energy harvesting circuits," IEEE Journal on Emerging and Selected Topics in Circuits and Systems, vol. 2, no. 1, pp. 24-33, 2012.

[4] K. A. Devi, S. Sadasivam, N. M. Din, and C. K. Chakrabarthy, "Design of a 377 Ω patch antenna for ambient RF energy harvesting at downlink frequency of GSM 900," in Proceedings of the 17th Asia Pacific Conference on Communications (APCC '11), pp. 492-495, Sabah, Malaysia, October 2011.

[5] R. Rahim, F. Malek, S. F. W. Anwar, S. L. S. Hassan, M. N. Junita, and H. F. Hassan, "A harmonic suppression circularly polarized patch antenna for an RF ambient energy harvesting system," in Proceedings of the IEEE Conference on Clean Energy and Technology (CEAT '13), pp. 33-37, IEEE, Lankgkawi, Malaysia, November 2013.

[6] M. Krakauskas, A. M. A. Sabaawi, and C. C. Tsimenidis, "Suspended patch microstrip antenna with cut rectangular slots for RF energy harvesting," in Proceedings of the 10th Loughborough Antennas and Propagation Conference (LAPC '14), pp. 304-307, Loughborough, UK, November 2014.

[7] Albert Sabban, " Low visibility Antennas for communication systems" , TAYLOR & FRANCIS group, 2015, USA.

[8] Albert Sabban, "Wideband RF Technologies and Antenna in Microwave Frequencies", Wiley Sons, July 2016, USA.

[9] Albert Sabban (2018). "Wearable Communication Systems and Antennas for Commercial, Sport, and Medical Applications". IET Publication, December 2018.

[10] Albert Sabban. Novel Wearable Antennas for Communication and Medical Systems; Taylor & Francis Group: Boca Raton, FL, USA, 2017.

Section 2

Green Programmable and Computing Systems

Chapter 2

Wideband Systems with Energy Harvesting Units for 5G, Medical and Computer Industry

Albert Sabban

Abstract

Demand for green energy is in tremendous growth in the last decade. The continuous growth in production of portable RF systems increase the consumption of batteries and electrical energy. Batteries and conventional electrical energy increase the environmental pollution. Compact wideband efficient antennas are crucial for energy harvesting commercial portable sensors and systems. Small antennas have low efficiency. The efficiency of 5G, IoT communication and energy harvesting systems may be improved by using wideband efficient antennas. Ultra-wideband portable harvesting systems are presented in this chapter. This chapter presents new Ultra-Wideband energy harvesting system and antennas in frequencies ranging from 0.15GHz to 18GHz. Three wideband antennas cover the frequency range from 0.15GHz to 18GHz. A wideband metamaterial antenna with metallic strips covers the frequency range from 0.15GHz to 0.42GHz. The antenna bandwidth is around 75% for VSWR better than 2.3:1. A wideband slot antenna covers the frequency range from 0.4GHz to 6.4GHz. A wideband fractal notch antenna covers the frequency range from 6GHz to 18GHz. Printed passive and active notch and slot antennas are compact, low cost and have low volume. The active antennas may be employed in energy harvesting portable systems. The antennas and the harvesting system components may be assembled on the same, printed board. The antennas bandwidth is from 75–200% for VSWR better than 3:1. The antennas gain is around 3 dBi with efficiency higher than 90%. The antennas electrical parameters were computed by using 3D electromagnetic software in free space and in vicinity of the human body. There is a good agreement between computed and measured results.

Keywords: green energy, energy harvesting, wideband antennas, metamaterial antenna, notch, slot antennas

1. Introduction

This chapter presents new Ultra-Wideband programmable communication systems with energy harvesting modules and efficient compact antennas in frequencies ranging from 0.15GHz to 18GHz. In last twenty years free space energy in the forms of light, heat, vibration, electromagnetic waves, muscle motion and other type of energy, is used to produce green energy. Methods to produce green electricity from these different types of energy sources have been presented and investigated [1–4].

Energy harvesting systems provide green energy and may eliminate the need to replace batteries every day and the usage of power cords. Batteries and cables waste pollute the environment. To use as much free space energy as possible it is important to harvest the electromagnetic power from wideband range of wireless communication systems. In these cases, we should use ultra-wideband antennas. Moreover, a programmable array with two to four antennas can harvest energy from 100 MHz to 18GHz. The energy harvesting antenna must satisfy the requirements related to the system application. Due to considerably low-power electromagnetic energy densities in free space, highly efficient antennas are significant. Patch, slot, and dipole antennas were employed to harvest electromagnetic energy [4–7]. Printed antennas are used in communication and medical system [4–25]. Wideband slot and notch antennas may be used in wideband harvesting energy systems. Slot and notch antennas have low volume, low weight, low cost, and are flexible. Moreover, a compact low-cost energy harvesting network and matching network may be produced by integrating the system RF components with the resonators on the same board. Printed compact antennas are widely presented in the literature in the last twenty years as referred in [7–25]. Human body effect on the electrical performance of wearable medical system and antennas at microwave frequencies should be considered. Electrical properties of human tissues have been presented in several papers such as [26–27]. Several wearable antennas were presented in books and papers in the last years as referred in [27–36]. Printed notch and slot antennas for harvesting energy applications are rarely presented in the literature. New ultra-wideband wearable antennas for 5G, IOT and medical RF systems with energy harvesting units are presented in this chapter.

2. Energy harvesting systems for 5G, IOT, medical and computer industry

In the last decade there is a significant increase in the amount of electromagnetic energy in the air. Almost every person has a cellular phone tablet and other communication devices. In electromagnetic energy harvesting systems, the electromagnetic waves propagating in free space may be received by harvesting antennas and converted to electric energy that is used to charge batteries and for other devices. The expected amount of radio wave in the air in 2019 was 33 Exa-bytes, EB, per month. However, the expected amount of radio wave in the air in 2025 is expected to be 164 Exa-bytes per month, see **Table 1**. **Table 1** presents the expected amount of radio wave in the air for 2G, 3G, 4G, and 5G networks. 5G is forecast to account for 45% of global mobile data traffic by 2025. Computations per KWh from 1985 to

Year	Total amount of radio wave in free space EB per month	Amount of radio wave in free space EB per month 2G, 3G, 4G	Amount of radio wave in free space EB per month 5G
2015	4.4	4.4	0
2017	11	11	0
2019	33	33	0
2021	60	50	10
2023	100	75	25
2025	164	85	79

Table 1.
Expected amount of radio wave in free space from 2015 up to 2025.

2018 are listed in **Table 2**. Energy sources used in harvesting systems are listed in **Table 3**. Communication services such as television, GSM, wireless local area networks, WLAN, and Wi-Fi covers the frequency range from 0.2 GHz to 5.4 GHz. Wireless communication systems operate in the frequencies from 700 MHz to 2700 MHz. Medical systems operate in the frequencies from 200 MHz to 1200 MHz. WLAN systems operate in the frequencies from 5400 MHz to 5900 MHz.

RF energy is inversely proportional to distance and therefore drops as the distance from a source is increased. Harvested power from RF energy sources is lower than 0.1 mW/cm^2. Electromagnetic energy harvesting system is shown in **Figure 1**. A programmable array with 3 to 4 antennas can harvest RF energy from 0.15GHz up to 18GHz. The received RF energy may be combined after transformed to DC power.

The RF energy harvesting system consists of antennas, matching and feed networks, rectifying circuit, and a rechargeable battery. The harvesting energy system operates as a dual mode RF harvesting system. The harvesting unit can be part of a medical, IOT, computer, and smartphone. The LNA DC bias voltages are supplied by the receiving system The Low Noise Amplifier is part of the. The energy coupled to the transmitting built in test, -20 dB, may be harvested and used to charge a battery. We can calculate the energy harvesting link budget by using Eq. (1), [8], if

Year	Computations per KWh (1E+09)
1985	50
1987	100
1992	1000
1997	10,000
2003	100,000
2008	1,000,000
2010	15,000,000
2012	20,000,000
2014	30,000,000
2016	40,000,000
2018	50,000,000

Table 2.
Computations per KWh from 1985 to 2018.

Energy source	Type	Efficiency	Estimated harvested power
Electromagnetic	0.3 – 5GHz Wi-Fi, WLAN	~50%	1 μW/cm^2 0.01 μW/cm^2
Wireless 1-3GHz	Wireless	~50%	0.1 up to 5 mW/cm^2
Light	Outdoor or Indoor	10 ~ 25%	150 mW/cm^2
Thermal	Human Industrial	~0.1% ~3%	60 μW/cm^2 ~1–10 mW/cm^2
Vibration	~Hz–human ~kHz–machines	20 ~ 50%	~4 μW/cm^3 ~800 μW/cm^3

Table 3.
Energy sources used in harvesting systems.

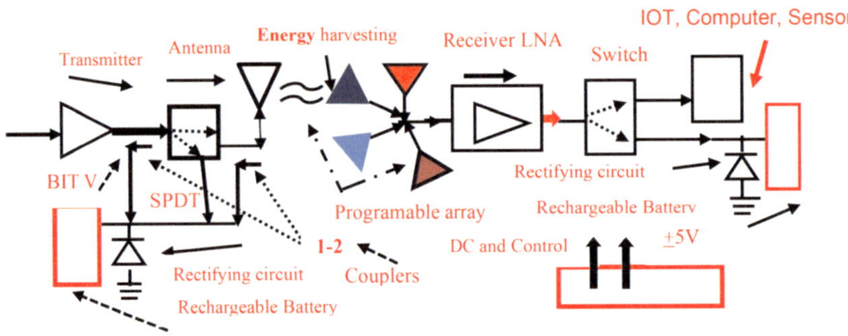

Figure 1.
Dual mode programmable Array energy harvesting system.

the antennas are matched and there are no losses in the medium. However, if the antennas are not matched and the RF energy propagates in a lossy media, we can calculate the energy harvesting link budget by using Eq. (2), [8]. Wireless smart phone using standard 802.11 can transmit up to 1Watt. Free Space Loss (Lp) represents propagation loss in free space. Losses due to attenuation in atmosphere, La = $e^{-\alpha r}$, should be accounted for in the transmission equation.

$$P_r = P_t G_t G_r \left(\frac{\lambda}{4\pi R}\right)^2 \tag{1}$$

$$P_r = P_t G_r G_t \left(\frac{\lambda}{4\pi r}\right)^2 \left(1 - |\Gamma_t|^2\right)\left(1 - |\Gamma_r|^2\right)\left|a_t a_r^*\right|^2 e^{-\alpha r} \tag{2}$$

The attenuation constant is α. Where, $L_p = \left(\frac{4\pi R}{\lambda}\right)^2$. The received power may be given as: $P_r = \frac{P_t G_t G_r}{L_p}$. Losses due to polarization mismatch, Lpol = $|a_t a_r|^2$, should also be accounted. Losses associates with receiving antenna, Lra, and with the receiver, Lr = $(1 - |\Gamma_r|^2)$, should be accounted in computation of transmission budget. Losses associates with the transmitting antenna as written as, Lta = $(1 - |\Gamma_t|^2)$. Where: Γ_r is the reflection coeficient of the receiving antenna.

$$P_r = \frac{P_t G_t G_r}{L_p L_a L_{ta} L_{ra} L_{pol} L_o L_r} \tag{3}$$

Γ_t is the reflection coeficient of the transmitting antenna.Pt = Pout/Lt, see Eq. (3), [8].

Pt = Transmitting antenna power. Lt = Loss between power source and antenna.

EIRP = Effective isotropic radiated power. Where, EIRP = Pt Gt, see Eq. (4), [8].

Where,

$$\begin{aligned} P_r &= \frac{P_t G_t G_r}{L_p L_a L_{ta} L_{ra} L_{pol} L_{other} L_r} \\ &= \frac{EIRP \text{ x } G_r}{L_p L_a L_{ta} L_{ra} L_{pol} L_{other} L_r} \\ &= \frac{P_{out} G_t G_r}{L_t L_p L_a L_{ta} L_{ra} L_{pol} L_{other} L_r} \end{aligned} \tag{4}$$

Gain, G, in dB; and L, Loss in dB is written as: $G = 10 \cdot \log\left(\frac{P_{out}}{P_{in}}\right)$ dB $L = 10 \log\left(\frac{P_{in}}{P_{out}}\right)$ dB.

The received power Pr in dBm may be calculated by using Eq. (5), [8]. The received power Pr is referred to as the "Carrier Power". Personal Computer Memory *Cards* using standard 802.11 can transmit up to 100 mW.

$$P_r = EIRP - L_{ta} - L_p - L_a - L_{pol} - L_{ra} - L_{other} + G_r - L_r \tag{5}$$

Figure 2 presents a wideband receiving system, 500 MHz -18GHz, with energy harvesting units.

Figure 3 presents a wideband transmitting system, 500 MHz to 18GHz, with energy harvesting units.

Three wideband antennas are employed in the receiving and transmitting systems presented in **Figures 2** and **3**. The first antenna covers frequencies from 150 MHz to 0.5GHz. The second antenna covers frequencies from 500 MHz to 6GHz. The third antenna covers frequencies from 6GHz to 18GHz. The receiving and transmitting systems can harvest energy from 150 MHz to 18GHz. Almost in every transmitting or receiving channel part of the energy, -10 dB to -20 dB, is coupled to a built-in test port. This RF energy can be transformed to DC energy and can be used to charge electrical devices. In the receiving channel the received RF energy from 150 MHz to 6GHz is transferred via a wideband combiner or SPDT to a second SPDT that is connected to the 6GHz to 18GHz antenna. The output of the second SPDT is connected to a third wideband SPDT via a wideband LNA. The LNA is part of the receiving channel. The output of the third SPDT transfer the receiving

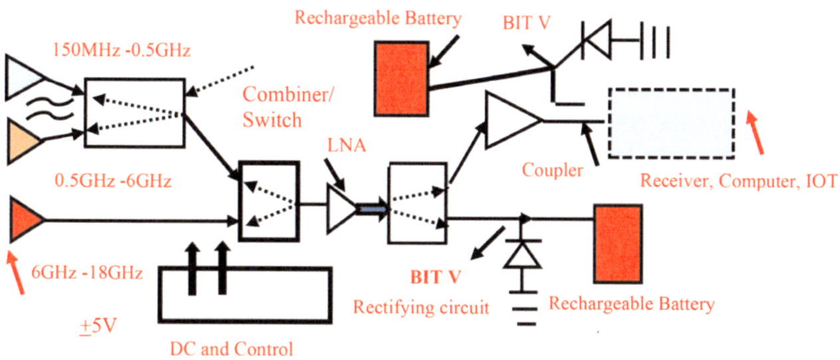

Figure 2.
Wideband harvesting receiving programmable Array system.

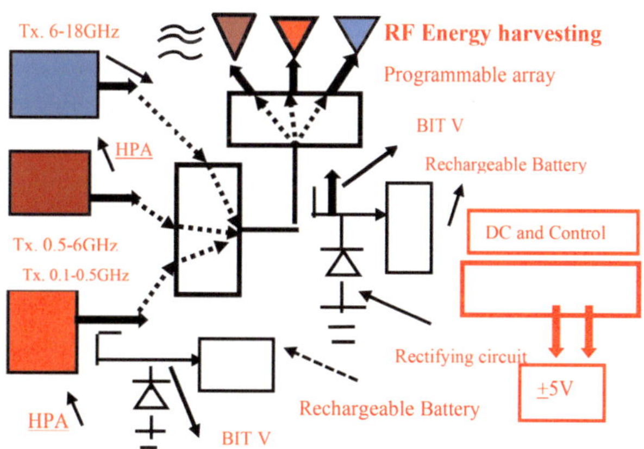

Figure 3.
Optimized harvesting transmitting programmable Array system.

RF energy to the receiving channel or to the harvesting system. A wideband amplifier transfers the received RF signal to the receiver. The received RF energy is coupled, -10 dB to -20 dB, is coupled to a Built in Test port. This RF energy is transformed to DC energy and may be used to charge electrical devices. A wideband switching matrix connects three power amplifiers to three transmitting antennas. The first antenna covers the frequency range from 150 MHz to 0.5GHz. The second antenna covers frequencies from 500 MHz to 6GHz. The third antenna covers frequencies from 6GHz to 18GHz. A DC unit supply the bias voltages to the amplifiers and control the switching matrix state. The harvesting system may consist one to three antennas according to the system requirements. The antennas may cover the frequency range from 150 MHz to 6.4GHz only.

3. Wideband 150 MHz to 0.4GHz, energy harvesting antenna

RF energy in the frequency range from150MHz to 500 MHz may be harvested by a wideband compact metamaterial antenna as presented in **Figure 4**. The antenna is a dual polarized printed dipole and slot antennas with Split Ring Resonators and metallic strips. The microstrip loaded dipole antenna with SRR provides horizontal polarization. The slot antenna provides vertical polarization. The dipole feed network is printed on the first layer. The radiating dipole with SRR is printed on the second layer. The thickness of each layer is 0.8 mm. The length of the antenna with SRR is 19.8 cm. The SRR ring width is 1.4 mm the spacing between the rings is 1.4 mm. The antennas have been analyzed by using electromagnetic software. The matching stubs and metallic strips locations and dimensions have been optimized to get the best VSWR results. The S11 parameter of the metamaterial antenna with metallic strips is presented in **Figure 5**. The antenna bandwidth is around 75% for VSWR better than 2.3:1.

The antenna radiation pattern is shown in **Figure 6**. The 3D computed radiation pattern is shown in **Figure 7**. Directivity and gain of the antenna with SRR are around 5dBi.

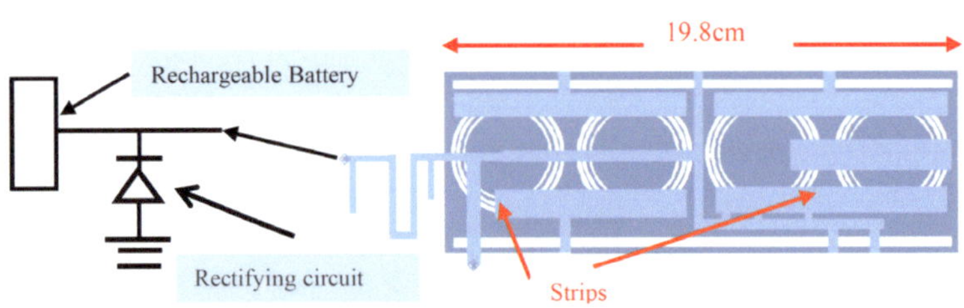

Figure 4.
Wideband antenna with SRR and metallic strips.

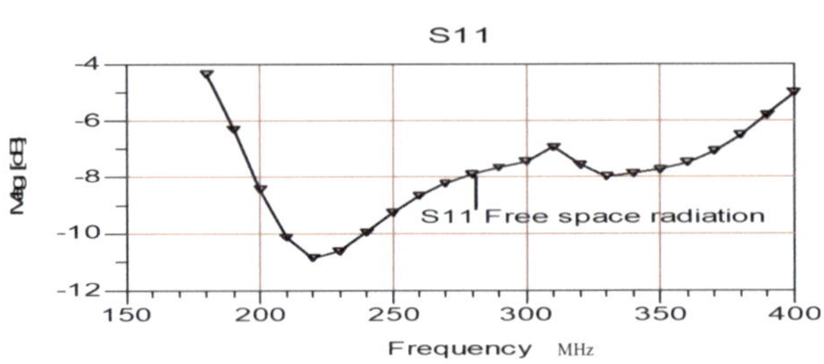

Figure 5.
S11 for antenna with SRR and metallic strips.

DOI: http://dx.doi.org/10.5772/intechopen.95879

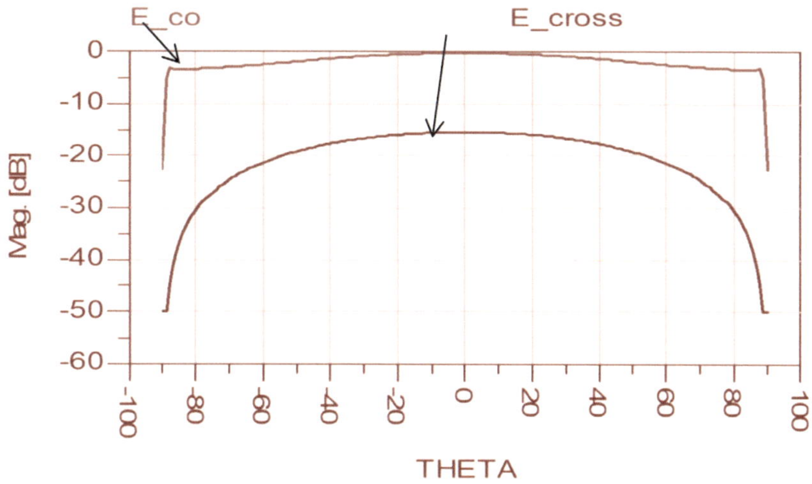

Figure 6.
Radiation pattern for antenna with SRR and metallic strips.

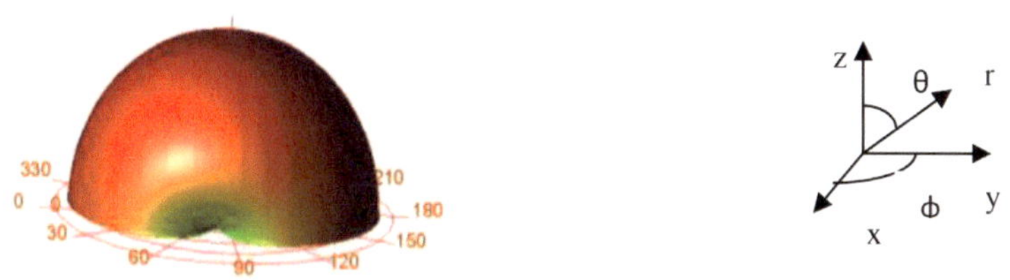

Figure 7.
3D radiation pattern for antenna with SRR.

4. Wideband 400 MHz to 6.4GHz, energy harvesting slot antenna

RF energy in the frequency range from 400 MHz to 6.4GHz may be harvested by a wideband compact T shape slot antenna as presented in **Figure 8**. The slot

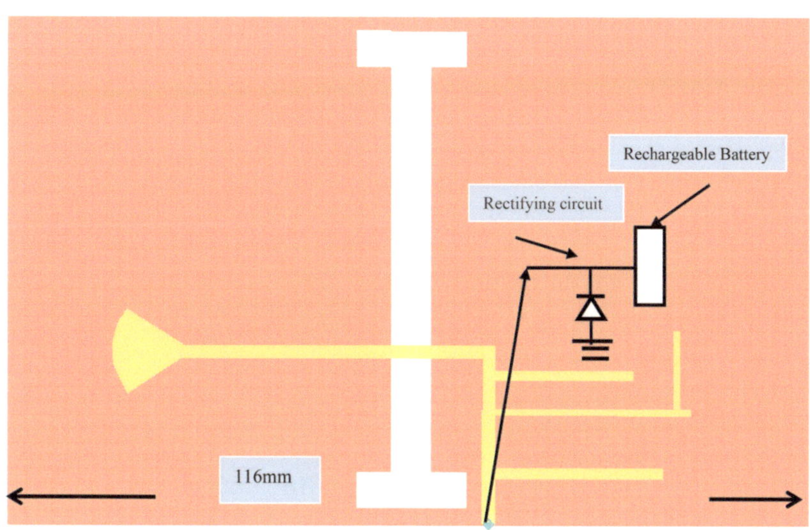

Figure 8.
A ultra-wide band T shape printed harvesting slot antenna.

antenna is printed on a dielectric substrate with dielectric constant of 2.2 and 1.2 mm thick. The dimensions of the slot antenna shown in **Figure 8** are 116x70x1.2 mm. The antenna electrical parameters were calculated and optimized

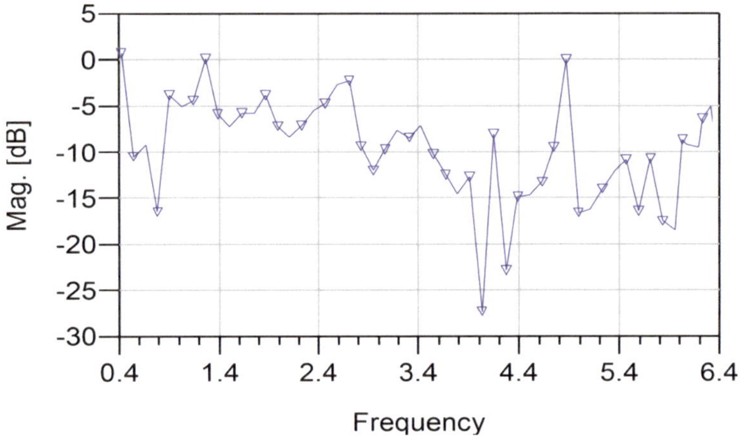

Figure 9.
S11 of a wide band printed T shape energy harvesting slot antenna.

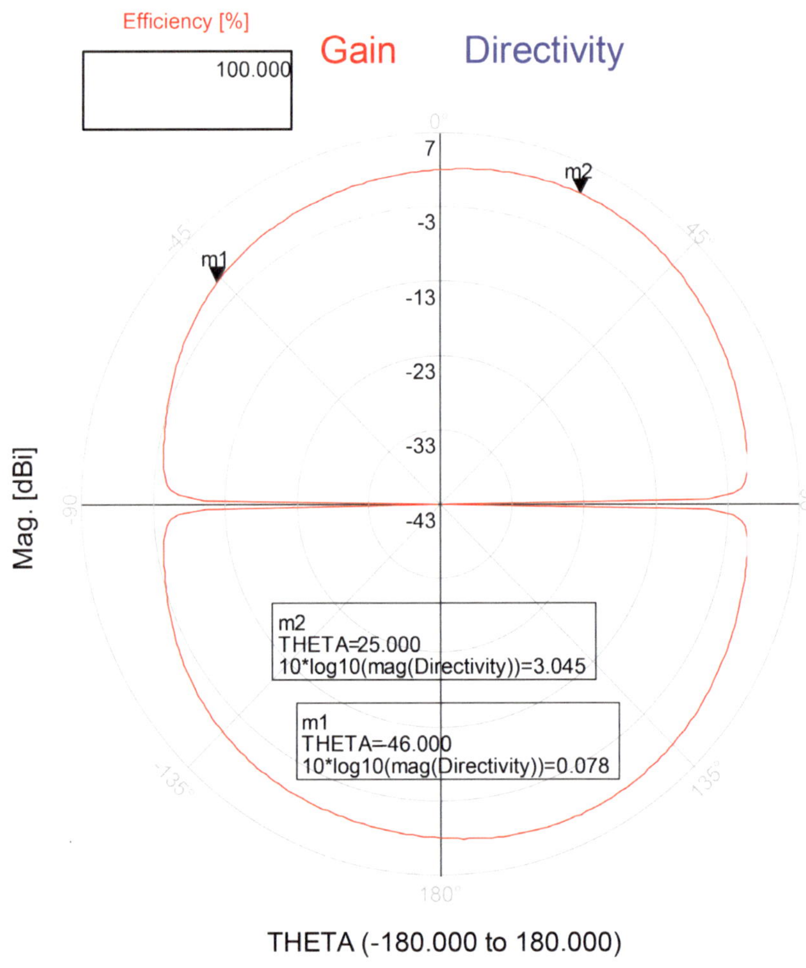

Figure 10.
Radiation pattern of a wide band wearable printed T shape slot antenna at 1.5GHz.

DOI: http://dx.doi.org/10.5772/intechopen.95879

by using ADS software [37]. The S11 parameter of the wideband T shape slot antenna is presented in **Figure 9**. The radiation pattern of the wideband T shape slot antenna at 1.5GHz is shown in **Figure 10**. The antenna gain is 3 dB.

5. New compact ultra-wideband harvesting notch antenna 5.8GHz to 18GHz

A wideband notch antenna with fractal structure was developed. The notch antenna is printed on a dielectric substrate with dielectric constant of 2.2 and 1.2 mm thick. The notch antenna is presented in **Figure 11**. The antenna volume is 11x7.7x1.2 mm. The antenna resonates from 5.8GHz to 18GHz. The antenna bandwidth is around 200% for VSWR better than 3.5:1, as presented in **Figure 12**. The notch antenna VSWR is better than 3:1 for more than 90% of the frequency range

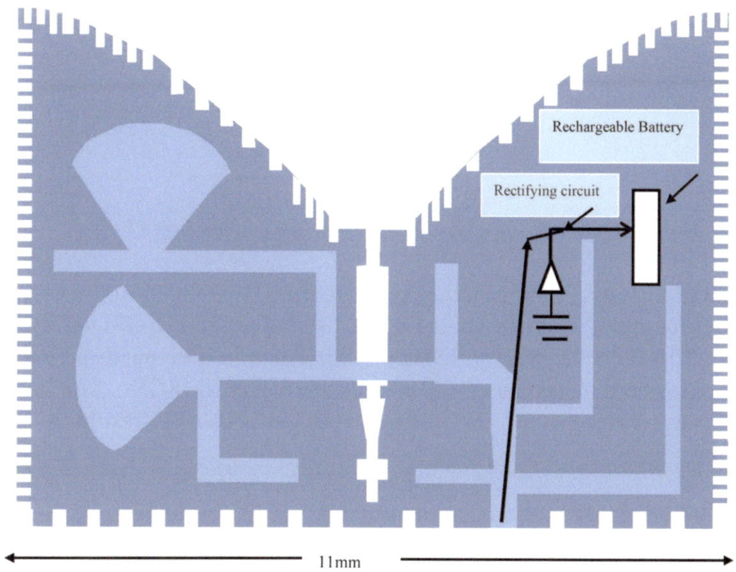

Figure 11.
A wideband 5.8GHz to 18GHz notch antenna with fractal structure.

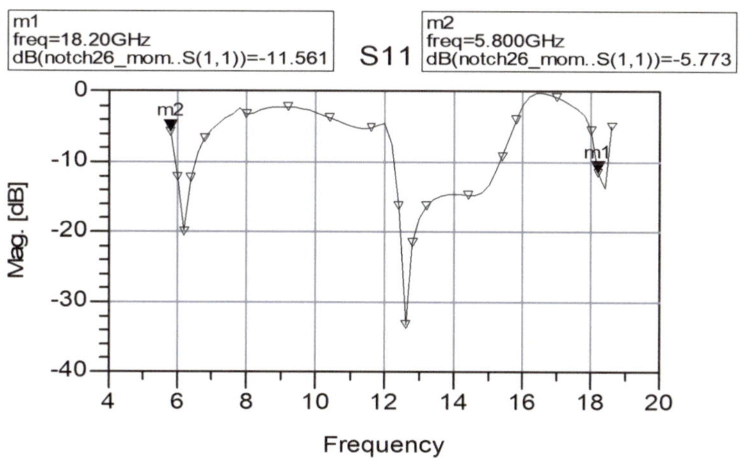

Figure 12.
A wideband 5.8GHz to 18GHz notch antenna with fractal structure, S11 results.

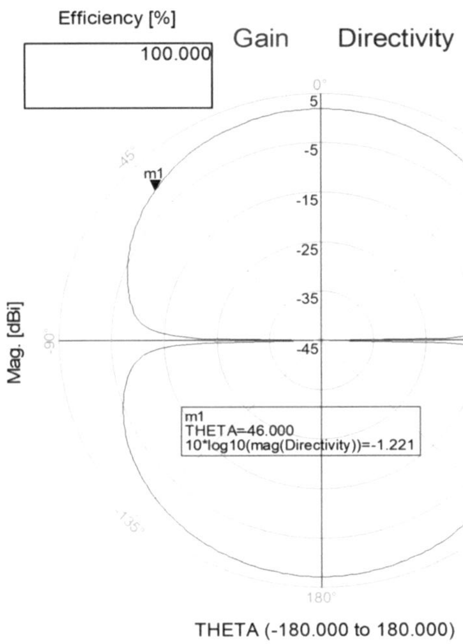

Figure 13.
Radiation pattern of the wideband notch antenna with fractal structure at 8GHz.

from 5.8GHz to 18GHz. The radiation pattern of the wideband notch antenna with fractal structure at 8GHz is shown in **Figure 13**. The antenna beam width at 6GHz is around 84°. The antenna gain, at 6GHz 3.5dBi. The antenna matching network was tuned and optimized to get better S11 results at 16GHz to 18GHz. The length and width of the stubs were tuned and optimized to get better S11 results at 16GHz to 18GHz.

6. Wideband energy harvesting systems and applications

As shown in **Figure 1** the programable array energy harvesting system consists of wideband programable antenna network, a rectifying circuit, and a rechargeable battery. A rectifier converts AC energy, alternating current AC, to DC energy (direct current DC). The popular rectifiers are half wave rectifier or full wave rectifier, [1–4]. A Half wave rectifier is presented in **Figure 14**. A half-wave rectifier conducts only during the positive voltage half cycle. It allows only one half of the electromagnetic waveform to pass through the load. The rectifier output DC voltage, V_{ODC}, is given in Eq. (6). The rectifier output voltage may be improved by connecting a capacitor in shunt to the resistor. The half wave rectifier with a capacitor is shown in **Figure 14**. *Vripple* is given in Eq. (7).

$$
\begin{aligned}
&V_{O,DC} = \frac{1}{2\pi}\int_0^{2\pi} V_O^{MAX}\sin(\omega t)d(\omega t);\ \ \omega = 2\pi f \\
&V_O = V_S - V_{DON} \approx V_S;\quad V_O{}^{MAX} = V_m \\
&V_{ODC} = V_m/\pi
\end{aligned}
\tag{6}
$$

$$
Vripple = Vr = Vmax - Vmin = \frac{VDC}{fCR} \tag{7}
$$

The time constant τ should be lower than T. Where,

$\tau = RC \ll T$. The half wave rectifier efficiency is 40.6%. Only 40.6% of the input AC power is converted into DC power. Where rf is the diode resistance is negligible as compared to R.

A Half wave bridge rectifier is shown in **Figure 15**. The bridge full wave rectifier circuit is used to convert AC energy to DC energy and for DC power suppliers. The bridge rectifier consists of four diodes D1 through D4, as shown in **Figure 15**. During the positive input half cycle, terminal A will be positive and terminal B will be negative. Diodes D1 and D2 will become forward biased and D3 and D4 will be reversed biased. The rectifier output DC voltage, $V_{ODC} = 2V_m/\pi$, The rectifier output voltage may be improved by connecting a capacitor in shunt to the resistor as shown in **Figure 15**.

The half wave rectifier efficiency is 81.2% as presented in Eq. (8). This means that only 81.2% of the input AC power is converted into DC power. The harvested voltage from the three antennas is V_{Total} as written in Eq. (9). The total DC energy is P_{Total} as written in Eq. (10). The actual harvested energy is P as written in Eq. (11).

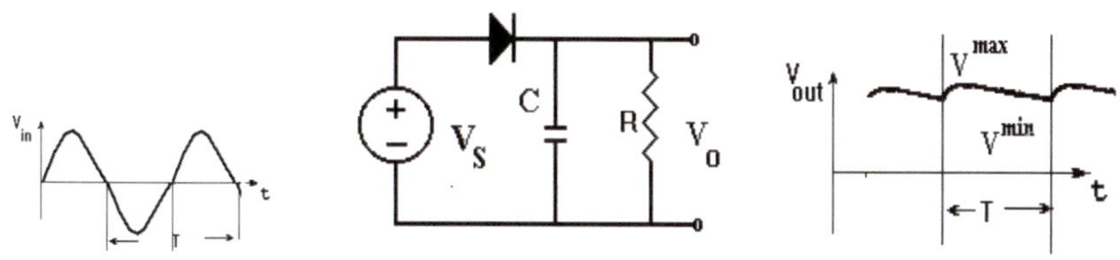

Figure 14.
Half wave voltage rectifier with a capacitor.

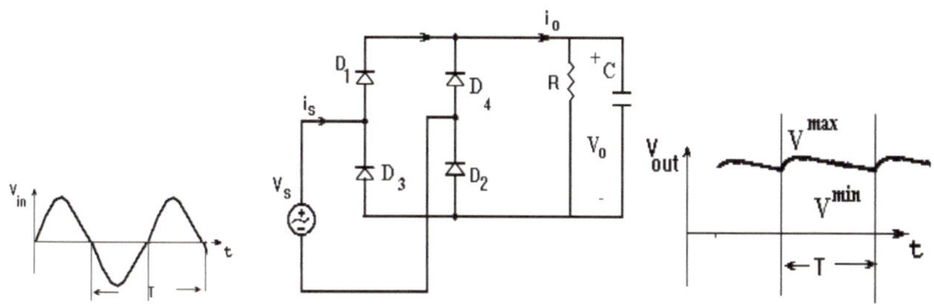

Figure 15.
Full wave bridge voltage rectifier with a capacitor.

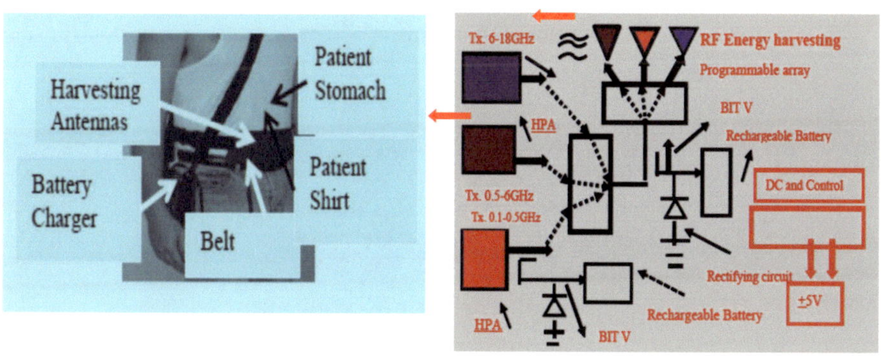

Figure 16.
Ultra-wideband RF system with energy harvesting modules for 5G and medical applications.

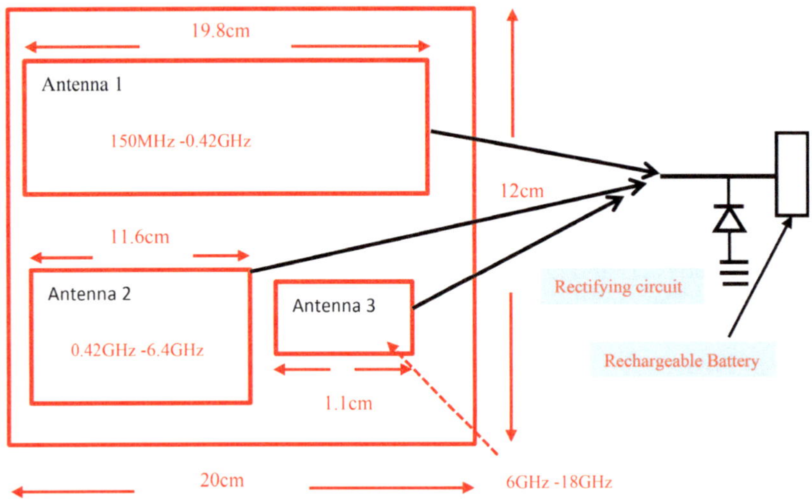

Figure 17.
Wideband RF energy harvesting panel.

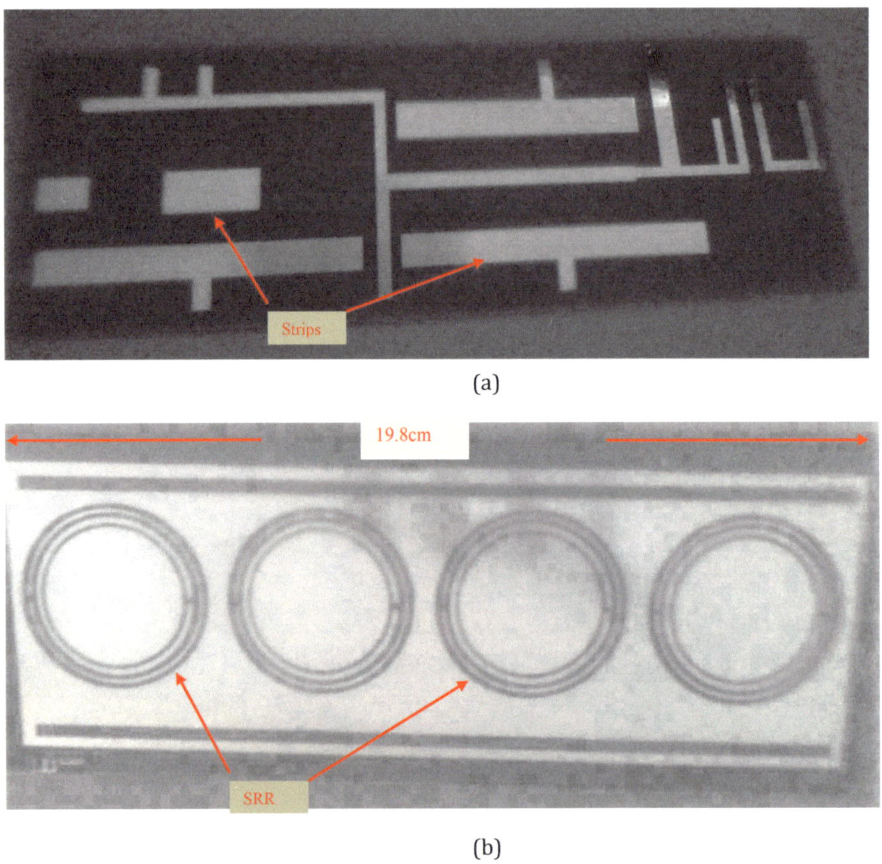

Figure 18.
Wideband energy harvesting metamaterial UHF antenna. (a) the antenna feed network and metallic strips. (b) Dual polarized UHF energy harvesting metamaterial antenna with four SRRs.

DOI: http://dx.doi.org/10.5772/intechopen.95879

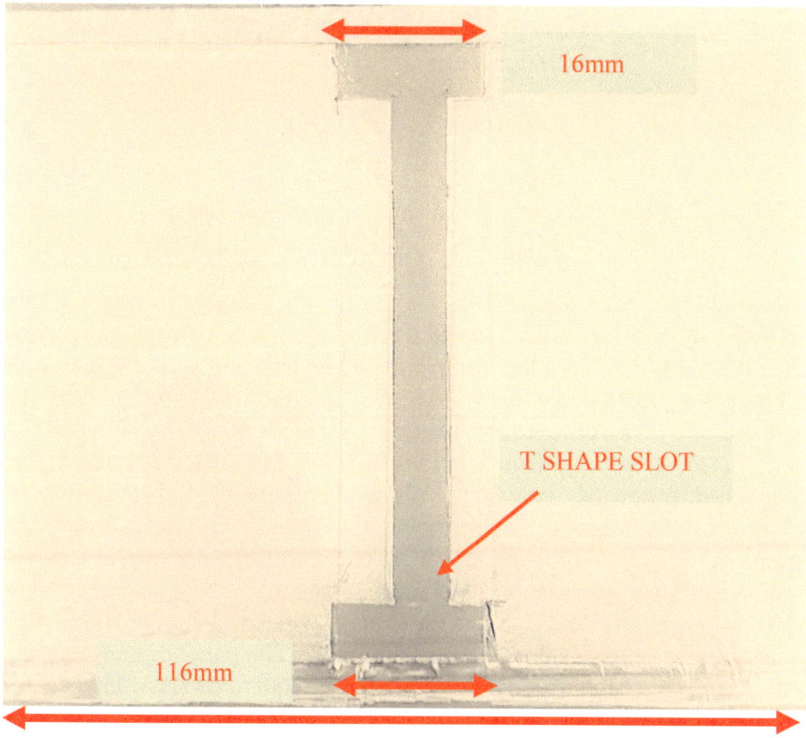

Figure 19.
A photo of wideband energy harvesting slot antenna, 400 MHz - 6 GHz.

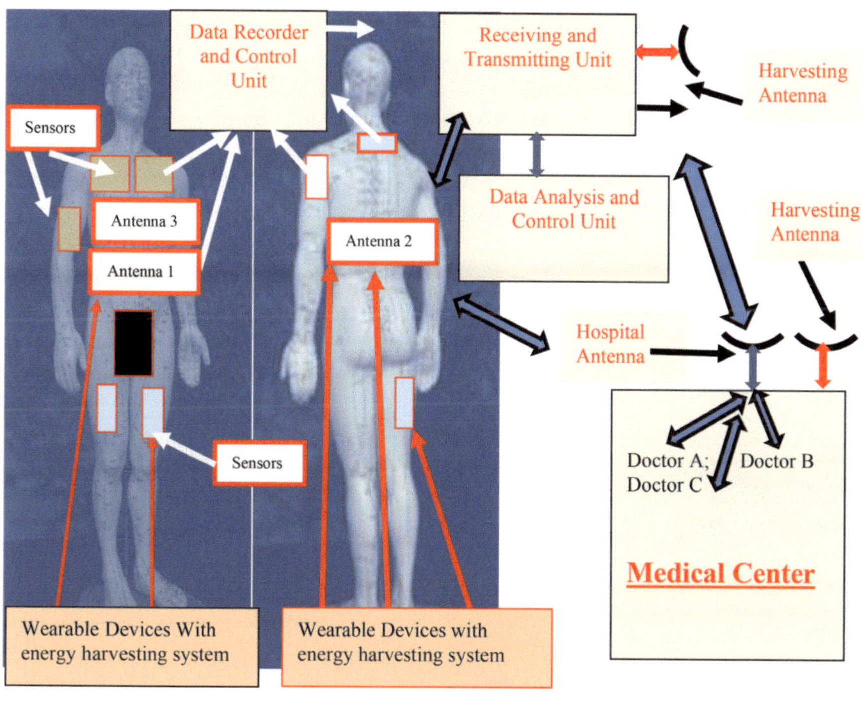

Figure 20.
Wearable body area network with energy harvesting system for medical applications.

$$\eta = \frac{\text{DCoutputpower}}{\text{ACinputpower}} = \frac{\left(\frac{2I_m}{\pi}\right)^2 R}{\left(\frac{I_m}{2}\right)^2 (R + rf)} 0.812 \quad (8)$$

$$V_{Total} = V_1 + V_2 + V_3 \quad (9)$$

$$P_{Total} = \frac{V_{Total}^2}{R} = I_m^2 R \quad (10)$$

$$P = 0.81 P_{Total} = 0.81 \frac{V_{Total}^2}{R} = 0.81 I_m^2 R \quad (11)$$

A Schottky diode may be used in the rectifier circuit. Schottky diodes are semiconductor diodes which has a low forward voltage drop and a very fast switching action. There is a small voltage drop across the diode terminals when current flows through the diode. The voltage drop of a Schottky diode is usually between 0.15 and 0.4 volts. This lower voltage drop provides better system efficiency and higher switching speed. A normal diode has a voltage drop between 0.6 to 1.7 volts. RF energy harvesting systems can be used to charge wearable devices. Ultra-Wideband RF System with energy harvesting modules for 5G and Medical applications is shown in **Figure 16**. **Figure 16** presents a wearable harvesting system with a wearable battery charger attached to a patient shirt. A wideband energy Harvesting panel with three antennas is presented in **Figure 17**. The panel dimensions are 20x12x0.02 cm. This panel can harvest energy in frequencies from 150 MHz to 18GHz as part of communication, medical, and IOT systems.

Figure 18 presents a photo of a wideband metamaterial antenna with metallic strips, 150 MHz –0.5GHz. **Figure 18a** presents the antenna feed network and the metallic strips. **Figure 18b** presents the dual polarized metamaterial antenna with four SRRs. **Figure 19** presents a photo of a wideband T shape slot antenna, 420 MHz –6.4GHz. These antennas can be used to harvest RF energy to charge wearable devices and sensors. The UHF dual polarized antenna can be attached to the patient stomach or back. The wideband T shape slot can be attached to the patient stomach or back. The wideband compact notch antenna can be also attached to the patient stomach or back. These antennas provide a wideband wearable communication system with a wideband RF harvesting system. Wearable Body Area Network with energy harvesting system for medical applications is presented in **Figure 20**.

A comparison of computed and measured results of compact wearable antennas for medical, 5G and IoT systems is listed in **Table 4**. Printed dipoles with and without SRR were presented in [9–10].

Antenna	Frequency (GHz)	Bandwidth %	VSWR	Computed Gain dBi	Measured Gain dBi
Printed dipole [9]	0.43	5–10	2:1	2–3	2–3
Dipole with SRR	0.4	8–12	2:1	5–7	5–7
Dipole (SRR and strips) [10]	0.14 to 0.42	UWB	2.5:1	5–7.5	5–7.5
Slot [10]	1 to 4	UWB	2:1	3	3
T shape slot [10]	0.4 to 6.4	UWB	3:1	3	3
Notch [10]	6 to 18	UWB	3:1	3	2–3

Table 4.
Comparison of electrical characteristics of energy harvesting antennas [9–10].

7. Conclusions

This chapter presents new Ultra-Wideband energy harvesting system and antennas in frequencies ranging from 0.15GHz to 18GHz. Three wideband antennas cover the frequency range from 0.15GHz to 18GHz. A wideband metamaterial antenna with metallic strips covers the frequency range from 0.15GHz to 0.42GHz. A wideband slot antenna covers the frequency range from 0.4GHz to 6.4GHz. A wideband fractal notch antenna covers the frequency range from 6GHz to 18GHz. The electromagnetic energy is converted to DC energy that may be employed to charge batteries, wearable medical devices, IOT, laptop batteries and commercial Body Area Networks, BANs. The harvesting energy system operates as a dual mode RF harvesting system. The harvesting unit can be part of a medical, IOT, computer, and smartphone. The notch and slot antennas were analyzed by using 3D full-wave software. Harvested power from RF transmitting links is usually lower than 0.1 μW/ cm^2. All antennas presented in this chapter can operate also as active antennas. Active antennas may improve the energy harvesting system efficiency. The wideband RF energy harvesting system consists of wideband antenna, DC and control unit, a rectifying circuit, and a rechargeable battery. The harvesting energy system operates as a Dual Mode Energy harvesting system. The active devices DC bias voltages are supplied by the communication system. The wideband programmable energy harvesting panel can harvest energy in frequencies from 150 MHz to 18GHz as part of communication, medical, and IOT systems. The antennas presented in this chapter provide a wideband wearable communication system with a wideband RF harvesting system. There is a good agreement between computed and measured results.

Author details

Albert Sabban[1,2]

1 Kinneret Academic College, Galilee, Israel

2 Ort Braude College, Karmiel, Israel

*Address all correspondence to: sabban@netvision.net.il

References

[1] J. A. Paradiso and T. Starner, "Energy scavenging for mobile and wireless electronics," IEEE Pervasive Computing, vol. 4, no. 1, pp. 18–27, 2005.

[2] C. R. Valenta and G. D. Durgin, "Harvesting wireless power: survey of energy-harvester conversion efficiency in far-field, wireless power transfer systems," IEEE Microwave Magazine, vol. 15, no. 4, pp. 108–120, 2014.

[3] P. Nintanavongsa, U. Muncuk, D. R. Lewis, and K. R. Chowdhury, "Design optimization and implementation for RF energy harvesting circuits," IEEE Journal on Emerging and Selected Topics in Circuits and Systems, vol. 2, no. 1, pp. 24–33, 2012.

[4] K. K. A. Devi, S. Sadasivam, N. M. Din, and C. K. Chakrabarthy, "Design of a 377 Ω patch antenna for ambient RF energy harvesting at downlink frequency of GSM 900," in Proceedings of the 17th Asia Pacific Conference on Communications (APCC '11), pp. 492–495, Sabah, Malaysia, October 2011.

[5] R. A. Rahim, F. Malek, S. F. W. Anwar, S. L. S. Hassan, M. N. Junita, and H. F. Hassan, "A harmonic suppression circularly polarized patch antenna for an RF ambient energy harvesting system," in Proceedings of the IEEE Conference on Clean Energy and Technology (CEAT '13), pp. 33–37, IEEE, Lankgkawi, Malaysia, November 2013.

[6] M. Krakauskas, A. M. A. Sabaawi, and C. C. Tsimenidis, "Suspended patch microstrip antenna with cut rectangular slots for RF energy harvesting," in Proceedings of the 10th Loughborough Antennas and Propagation Conference (LAPC '14), pp. 304–307, Loughborough, UK, November 2014.

[7] Albert Sabban, "Low visibility Antennas for communication systems" , TAYLOR & FRANCIS group, 2015, USA.

[8] Albert Sabban (2018). "Wearable Communication Systems and Antennas for Commercial, Sport, and Medical Applications". IET Publication, December 2018.

[9] Albert Sabban, "Wideband RF Technologies and Antenna in Microwave Frequencies", Wiley Sons, July 2016, USA.

[10] Albert Sabban. *Novel Wearable Antennas for Communication and Medical Systems*; Taylor & Francis Group: Boca Raton, FL, USA, 2017.

[11] A. Sabban and K.C. Gupta, "Characterization of Radiation Loss from Microstrip Discontinuities Using a Multiport Network Modeling Approach", I.E.E.E Trans. on M.T.T, Vol. 39,No. 4,April 1991, pp. 705–712.

[12] A. Sabban,"A New Wideband Stacked Microstrip Antenna", I.E.E.E Antenna and Propagation Symp., Houston, Texas, U.S.A, June 1983.

[13] A. Sabban, E. Navon "A MM-Waves Microstrip Antenna Array", I.E.E.E Symposium, Tel-Aviv, March 1983.

[14] Balanis, C.A. "Antenna Theory: Analysis and Design", 2nd Edition. Wiley, 1996

[15] A. Sabban, PhD Thesis, Multiport Network Model for Evaluating Radiation Loss and Coupling Among Discontinuities in Microstrip Circuits, University of Colorado at Boulder, January 1991.

[16] Inventors: Albert Sabban, *Microstrip antenna arrays*, U.S. Patent US 1986/ 4,623,893, 1986.

[17] A. Sabban, "Wideband Microstrip Antenna Arrays", I.E.E.E Antenna and

Propagation Symposium MELCOM, Tel-Aviv June 1981.

[18] K. Fujimoto, J.R. James, Editors, Mobile Antenna Systems Handbook, Artech House, Boston USA, 1994.

[19] A. Sabban, "New Wideband Notch Antennas for Communication Systems", Wireless Engineering and Technology Journal, pp. 75–82, April 2016.

[20] Inventors: Albert Sabban, Dual polarized dipole wearable antenna, U.S Patent number: 8203497, June 19, 2012, USA.

[21] A. Sabban (2012) Wideband Tunable Printed Antennas for Medical Applications. I.E.E.E Antenna and Propagation Symp., Chicago IL. ,U.S.A, July 2012.

[22] A. Sabban, "New Wideband printed Antennas For Medical Applications" I.E. E.E Trans. on Antennas and Propagation, January 2013, Vol. 61, No. 1, pp. 84–91.

[23] A. Sabban, "Comprehensive Study of Printed Antennas on Human Body for Medical Applications" International Journal of Advance in Medical Science (AMS), Vol. 1, pp. 1–10, February 2013.

[24] R. Kastner, E. Heyman, A. Sabban, "Spectral Domain Iterative Analysis of Single and Double-Layered Microstrip Antennas Using the Conjugate Gradient Algorithm", I.E.E.E Trans. on Antennas and Propagation, Vol. 36, No. 9, Sept. 1988, pp. 1204–1212.

[25] A. Sabban, "Microstrip Antenna Arrays", Microstrip Antennas, Nasimuddin Nasimuddin (Ed.), ISBN: 978–953–307-247-0, InTech, http://www.intechopen.com/articles/show/title/microstrip-antenna-arrays , pp..361–384, 2011.

[26] Lawrence C. Chirwa*, Paul A. Hammond, Scott Roy, and David R. S. Cumming, "Electromagnetic Radiation from Ingested Sources in the Human Intestine between 150 MHz and 1.2 GHz", IEEE Transaction on Biomedical eng., VOL. 50, NO. 4, April 2003, pp 484–492.

[27] D.Werber, A. Schwentner, E. M. Biebl,"Investigation of RF transmission properties of human tissues", Adv. Radio Sci., 4, 357–360, 2006.

[28] Gupta, B., Sankaralingam S., Dhar, S.,"Development of wearable and implantable antennas in the last decade", Microwave Symposium (MMS), 2010 Mediterranean 2010 , Page(s): 251–267.

[29] Thalmann T., Popovic Z., Notaros B. M, Mosig, J.R.," Investigation and desig n of a multi-band wearable antenna", 3rd European Conference on Antennas a nd Propagation, EuCAP 2009. Pp. 462–465.

[30] Salonen, P., Rahmat-Samii, Y., Kivikoski, M.," Wearable antennas in the vicinity of human body", IEEE Ante nnas and Propagation Society Interna tional Symposium, 2004. Vol.1 pp. 467–470.

[31] Kellomaki T., Heikkinen J., Kivikoski, M., " Wearable antennas for FM reception", First European Conference on Antennas and Propaga tion, EuCAP 2006 , pp. 1-6.

[32] A. Sabban, "Comprehensive Study of Printed Antennas on Human Body for Medical Applications" International Journal of Advance in Medical Science (AMS), Vol. 1, pp. 1–10, February 2013.

[33] A. Sabban, "Wideband printed antennas for medical applications" APMC 2009 Conference, Singapore, Dec. 2009.

[34] A. Alomainy, A. Sani et all "Transient Characteristics of Wearable Antennas and Radio Propagation

Channels for Ultrawideband B ody-centric Wireless Communication", I.E.E.E Trans. on Antennas and Propagation, Vol. 57, No. 4, April 2009, pp. 875–884.

[35] M. Klemm and G. Troester, "Textile UWB antenna for Wireless Body Area Networks", I.E.E.E Trans. on Antennas and Propagation, Vol. 54, No. 11, Nov. 2006, pp. 3192–3197.

[36] P. M. Izdebski, H. Rajagoplan and Y. Rahmat-Sami, "Conformal Ingestible Capsule Antenna: A Novel Chandelier Meandered Design", I.E.E.E Trans. on Antennas and Propagation, Vol. 57, No. 4, April 2009, pp. 900–909.

[37] ADS software, Agilent, http://www.home.agilent.com/agilent/product.jspx?cc=IL&lc=eng&ckey=1297113&nid=34346.0.00&id=1297113

Chapter 3

Internet Connectivity in Building Interiors: Architecture and Sustainability Considerations

Zebun Nasreen Ahmed and Mohammad Tanvir Kawser

Abstract

This paper addresses the issue of connectivity and its relationship to the architecture of spaces. The internet of things and wifi technology is coming into the forefront of lifestyles in present times. We tend to spend much of our life activities using devices that need internet connectivity, wifi coverage, both indoors and outdoors. The technology behind it is also in the process of shifting gears, from 3G to 4G to 5G soon. However questions arise: How suitable are existing buildings for the sorts of internet technology that are predicted for the future? And how can the coverage and connectivity issue be addressed to give greater efficiency to new buildings? Will we require greater open layouts that give more visual connectivity within interior spaces to permit efficient connectivity between routers, boosters and appliances? The expertise of both Architects who design spaces, as well as RF Engineers who are responsible for setting up the internet systems within the spaces are involved in efficiently handling these queries. The paper incorporates recent studies have revealed that certain building materials may be more suited to Wifi penetration, while others act in a more opaque way. Architects are the main decision-makers regarding the final choice of such questions, but they have very little grounding in the technicalities involved related to connectivity issues. It is important to have them focus on these questions and seek answers through multi-disciplinary forums, thus also to deepen collaboration between the related professionals, in order to create more Wifi connectivity in building interiors. From the user perspective, this is a vital need, as the user of today spends enormous amounts of time at these electronic, computing devices that require uninterrupted and quality connectivity. It thus also becomes a question of sustainability.

Keywords: wifi connectivity, architecture, propagation losses, sustainability, multi-disciplinary approaches

1. Introduction

This book discusses issues about emerging technologies and the computing industry, and the focus of this chapter is on internet connectivity within building interiors. The chapter begins with a background on the needs for such connectivity, why Architects need to be concerned with such technical matters, and the need for a multi-disciplinary approach to the issue. It then elaborates on some of the key concepts of connectivity that affect the architecture of spaces, extracting from a study done by the authors, to determine building interior components that affect connectivity, impacting propagation of radio waves. The results of the study point to certain interior features and their design, to increase efficiency of the system, and minimize propagation losses.

In the discussion, conflicts, between best design scenarios focused on the connectivity issue, and satisfaction of other demands on internal spaces, are revealed. The field is still very new, as development in internet connection and propagation is progressing exponentially, with improvements taking place almost continuously. And if there is a gap in the transfer of knowledge, regarding what is required for good propagation, and what is being designed to house these systems, there is likely to be a clash of intentions, resulting in inefficient systems, and reduced user satisfaction. It is the intention of the authors, through this chapter to bring awareness to these inter-related issues, and to the need for collaboration, between the different professionals designing and implementing the spaces, on the one hand, and the systems, on the other.

In a nutshell, the chapter addresses the issue of the Internet of Things (IoT), wireless connectivity and related issues, as affected by the architecture of built spaces. Based on a recent study completed at BUET on the subject, and on related research, the issues are looked at broadly, from an architectural point of view, and steps that need to be addressed immediately to resolve these, are touched on, with suggested guidelines.

2. Lifestyle changes and need for connectivity

There has been a major shift in lifestyles caused by the Fourth Industrial Revolution (4IR). This revolution can be defined as a new chapter in human development, built on the previous three industrial revolutions that had shaped life since the mid-eighteenth century onwards: the first developing from water/steam power, the second from electric power, just before the beginning of the twentieth century, and the third building on developments in electronics and information technology, in the second half of the twentieth century [1].

This fourth revolution, begun at the turn of this century, has been enabled by extraordinary technological advances, superseding the three preceding development phases, and here we see a fusion of technology, which blurs the lines between the physical, digital and biological spheres. This affects the way that the world is developing in three ways – the speed/pace (exponential) at which changes take place, the scope of the changes (due to unprecedented processing power, access to knowledge, artificial intelligence, blurring of boundaries, digital, nanotechnology, material science, etc), and the impact of the systems (on production, management and governance). And from the point of view of life, this fourth industrial revolution has completely changed the lifestyle of humans. Those who can take advantage of this revolution are the ones that will see progress, others will lag behind. And the scale of the gap will increase incrementally, given the tremendous force and interconnectivity of the phenomenon.

The 4IR has ensured the passage of the internet of things and wireless technology into the forefront of life activities nowadays. The profusion of internet and information technology has led people to spend more and more time at their computers, smart devices and cell phones, spending their work time using these, and their leisure time enjoying them. Smart devices are used for functions ranging from communication, both officially and socially, to shopping, internationally and locally, to virtual meetings and social media, and has likewise extended to virtual education and training. This changed behavior has led the media to refer to today's youth as the 'indoor' generation, and research finds that around

90 percent of people spend close to 22 hours indoors each day in the developed world [2], at the present time.

Throughout history, people have tried to adapt Indoor conditions to suit their needs, and based them on their available resources and know-how. But the needs, as well as available knowledge base, have now changed very rapidly, and it is becoming important to examine how suited today's built spaces are, to accommodate these new needs, given the technology available to us. There are two major avenues of investigation here: whether the buildings allow proper conditions needed for such exchanges and activities, and whether the buildings can protect humans from any adverse effects of these exchanges.

Moreover, the technology behind the internet of things, is also in the process of shifting gears, from 3G to the present 4G, which is being upgraded to 5G soon. Up until now, this domain has been dominated only by RF (Radio Frequency) engineers, those who focus their expertise on the propagation of radio waves, and on their efficient and seamless transmission. RF Engineers work on improvement of the performance of wireless communication, by ensuring good RF signal strength at the receiver [3]. The need for Architects to be involved in these considerations is paramount at the moment, as it is they who are primarily responsible for designing and shaping the built environment, where all this propagation is to take place. They are the ones who will be ascertaining the arrival and receipt of any propagation, and thereby the success of such connectivity issues.

3. The building Internet of things (BIoT)

The Internet of Things (IoT) is expanding exponentially, with improvements in connectivity and the related technology. It is now an important instrument, in the process used to communicate with society, and is thus increasing the participatory approach to a myriad of things. This phenomenon is driving much of human activity, and is one of the most compelling factors that are making humans into sedentary beings, focussed solely on indoor living. When considered within buildings themselves, this phenomenon is termed BIoT, or Building Internet of Things (BIoT).

Another important innovation of recent times, is related to connectivity of equipment within 'smart' buildings, i.e. buildings having automated solutions for security, safety, energy management, comfort, entertainment, health, and so forth, all dependent on machine to machine (M2M) communication. This too is now becoming an integral part of BIoT and daily living, and occupants now depend on smart solutions within buildings, to ensure much that was previously under the domain of behavioral response, like adjusting window openings for thermal comfort, or controlling lighting level manually or using curtains and so on. In a smart building, service robots, and other devices for desired control, can be programmed to perform many of these activities, using 4G or 5G platforms, as indicated in **Figure 1** [4].

Recent studies have revealed that certain building materials and their properties may be more suited to wireless signal penetration, while others are relatively opaque. Since Architects are the main decision-makers, regarding the final choice of such questions as above, while they have very little grounding in the technicalities involved, it is important to have them reflect on such questions, and seek answers through multi-disciplinary forums, thus also to deepen collaboration between the related professionals, in order to create more wireless connectivity in building interiors. From the user perspective, this is a vital need, as the user of today spends enormous amounts of time at these electronic, computing devices that require

Figure 1.
Automated devices using 4G or 5G platform in smart buildings [4].

uninterrupted and quality connectivity. It is seen, therefore, that the performance of wireless communication is finding more direct bearing on the satisfaction level of users of the building, contributing thus to both their performance, as well as user comfort, above and over issues of comfort in terms of other environmental variables. It is becoming increasingly important now, to initiate studies on this aspect of comfort, as an added dimension of evaluating building performance, in a way that was in the past studied with reference to thermal and visual comfort issues.

4. The question of sustainability

In addition to the above considerations, any building activity or development today must be sustainable, and the business-as-usual model comes with inherent risks to the planet and continued human existence. This requires us to look at the issue in several separate dimensions. As buildings are one of the highest consumers of electricity, and Architects are largely responsible for their design, it is imperative that they be conversant with all the pathways in which energy efficiency can be achieved in buildings. Good internet connectivity within buildings too, is directly affected by the design of buildings. It, therefore, becomes incumbent to look at whether the satisfaction of these two requirements, for connectivity, and for efficiency, cause contradictory physical manifestations in building design.

Sustainability, considering the internet issue, can thus be approached, through an examination of connectivity needs, and any potential conflicts that they present, with other human and environment related issues. This involves studying the main linkages, synergies, and trade-offs with other issues, e.g. of health, comfort and productivity, bio-diversity, etc. As the issue of sustainability is intrinsically related to acceptance by user groups, it is also important to examine connectivity "solutions" available to the public and private sectors, their desirability, acceptance, affordability and other such factors, so that whatever solution is adopted for a building is inclusive, does not increase divides between different groups. This human factor is also one of the key indicators of sustainability.

5. SDGs, internet connectivity and architecture

A short discussion on the UN sustainable development goals (SDGs) set up in 2015, and the intrinsic interlink with the world-wide-web and internet connectivity is given below. If the spaces that serve such connectivity are not designed with seamless connectivity in view, there is no doubt that it will not be possible to achieve these goals. SDGs require that governments, private sector, civil society and citizens should partner together, aiming for a better planet for future generations [5]. And the road map to achieve this, banks largely on incorporating creativity, know-how, technology and financial resources available.

Goals 1 and 2 relate to deprived populations; no poverty and zero hunger. Much of poverty arises due to lack of opportunities, which is a direct result of not having proper access to available options of employment, business outlets, and so on. Clearly, these goals are interconnected, and cannot be tackled in isolation. Information exchange is vital, populations require to be connected through cell phones, etc. Goals 3 and 4, regarding health and education, are likewise related to connectivity. Proper monitoring, access, etc. are all dependent, on how easily information can be transferred from place to place, in order to stem problems at source. The internet is vital for present-day education, which can serve to be a tool, tackling the lack of libraries and information. Goal 8, decent work and economic growth, and Goal 9, industry, innovation and infrastructure, and again depend on increased connectivity, and the related issue of spaces and their design.

Goal 11 is directly about cities and thus architecture, and the way the built environment impacts our planet. The design of cities and urban areas affects the passage of internet connectivity, due to building positioning, spacing, material use and so forth, and thus decisions that impact these, are the direct concern of all the related professionals being discussed here.

Goal 12 relates to responsible consumption and production, and is also heavily dependent on internet availability. Connectivity within groups is vital to any business activity, therefore, the spaces where people produce, i.e. industries, and their design are vital in meeting this goal. Consumption is part and parcel of everyday living, and sustainable practices call for reduction, reuse and recycling, i.e. circular economies. Again connectivity is vital, to ensure that the system works properly and promptly, to avoid clogs and blockages within the flow.

Goal 13, is to control climate change. This is all pervasive, requiring attention to the design of built spaces, which is largely responsible for the carbon footprint and the consumption of resources on our planet. Buildings designed to conserve energy and use natural resources are important, just as are automated systems within buildings, which can monitor variables, and largely offset much waste of energy, and maintain the efficiency of systems. Such systems again cannot function without internet connectivity.

Goals 16 and 17 relate to people's issues and their participation. Connectivity and platforms for voicing opinions are imperative to achieve these, making the internet and spaces of use directly related to their success.

The above points have been raised, to underline the importance of treating the issue of connectivity as a key part of architecture considerations, and to the attainment of many of the SDGs. The need for the professionals to collaborate in the design of buildings, and spaces in between, is clear from these parallel associations.

6. Research relating architecture with RF matters

Architects, through their training, look at challenges existing in a given situation, and on means to address them through holistic solutions. Till recently, environmental forces were largely natural, but since the age of the information technology, the profusion of high frequency propagation has introduced a new dimension to ensure proper functioning, and with it are related health issues. The problem has been under the sole jurisdiction of Radio Frequency (RF) Engineers, who specialize in devices and transmission using and operating on radio waves, i.e. wireless devices like mobiles, which are largely becoming the centre of existence of modern lifestyles. But research shows that the efficiency of RF transmissions and reception, seems to be intrinsically associated with the design of spaces. It is therefore becoming imperative that Architects work hand-in-hand with RF Engineers, to create environments that are conducive to proper utilization of available RF regimes, to ease unnecessary losses, which can result in poor connectivity and energy inefficiency. This section summarizes a study done at BUET [6] and its follow up, to investigate the issue of internet connectivity and its connection to architecture.

6.1 Options for wireless service

The first point highlighted in the study is that building occupants are now demanding increasingly higher data rate, as new applications are emerging. Examples of these are: Augmented Reality (AR), Virtual Reality (VR), Mixed Reality (MR), and Extended Reality (XR). The global mobile data traffic is expected to increase from 19.01 exabytes per month in 2018 to 77.5 exabytes per month by 2022, at an annual growth rate of 46 percent [7].

The wireless connectivity in the building can be supported by many different technologies, as shown below.

1. Long Range High/Moderate Power: 3G/4G/5G cellular system.
    - For IoT devices, Narrowband-IoT (NB-IoT) or Extended Coverage GSM (EC-GSM) can be used.
    - In-Building Solutions (IBS): femtocells, picocells and Distributed Antenna System (DAS)
2. *Short Range High Speed:* Wi-Fi. For IoT devices, a version, called Wi-Fi HaLow (IEEE 802.11ah), with lower power consumption, has been introduced.
3. *Short Range Moderate Speed:* Wireless Personal Area Network (WPAN)
    - Bluetooth and BLE (Bluetooth Low Energy): Originally, standardized as IEEE 802.15.1 for operation in ISM radio bands.
    - ZigBee: Standardized as IEEE 802.15.4 for operation in ISM radio bands.
    - Thread: An IPv6-based, low-power mesh networking technology for IoT products.
    - Z-Wave: A low-power mesh networking technology, primarily used for home automation.
4. Long *Range Low Power:* Low Power Wide Area Networks (LPWAN)
    - Sigfox: A global cellular based network operator that supports IoT products at low-power.

- LoRa: A cellular based technology that supports IoT products at low-power.
- Waviot: A wireless technology that uses low-power and primarily, supports electricity meters and water meters.
- Ingenu: A provider of wireless networks that supports IoT products at low-power.
- Weightless: A set of open wireless technology standards that supports IoT products at low-power.

The chosen option for any particular building, will be affected by the geometry of the space, and its material quality. Therefore, contributions from architects can ease the job, towards ensuring proper wireless coverage and connectivity. Architects and RF engineers can complement each other, to address the growing challenge better. Green architecture considerations, apart from the physical features of the built spaces, can incorporate suggestions for various other measures, for example, solar panels, thermal mass building construction, green materials, including wood, stone, or earth, recycled waste materials, and so forth.

Architectural intervention can improve wireless signal coverage, by ensuring maximum signal power, minimizing interference to its path. Such intervention is significant, since the building will undoubtedly last much more than a few decades, and should thus be ready for the rapid changes that are predicted in the wireless support arena. However, there has not been enough engagement of architects in this area so far, and mostly, analytical discussions have been made in this regard [8], restricted within RF engineering circles, often beyond the knowledge of Architects.

6.2 Shifting gears: from 4G to 5G

4G has changed the life of people, but 5G is set to change society in its entirety. While the key focus of the developers for generations up to 4G, has been to improve data rate support, 5G has an additional focus, which is to support numerous use cases, with diverse technological requirements. These use cases are categorized with three basic types of requirements as shown below [9], all of which are seeing increasing applicability in new urban paradigms:

1. *Enhanced Mobile Broadband (eMBB):* The basic requirement of the eMBB use cases is very high data rate and its examples are, live HD videos, VR, and AR.
2. *Massive Machine-Type Communications (mMTC):* The basic requirement of the mMTC use cases is massive density of user devices, with each requiring very low data rates. Its example is Internet of Things (IoT) devices.
3. *Ultra-reliable low-latency communication (URLLC):* The basic requirements of the URLLC use cases are very low latency and very high reliability. Its examples are vehicular communication and automation in the industries.

The categorization of various use cases towards the three classes for 5G is illustrated in **Figure 2** [9].

Another complication in wireless connectivity, arises from the rapid changes taking place in the related technology. Due to the exponential growth of internet use, the lower frequencies of propagation are getting saturated. Service providers have incrementally shifted from 2G to 3G to 4G in the space of fewer than

Figure 2.
Categorization of various use cases towards the three classes for 5G [9].

2 decades, and are now geared to shift to 5G coverage. So cellular operation, functioning at frequencies below 6 GHz, were suitable for systems designed for propagations up to 4G. However, the need for 5G cellular systems is to encompass much higher frequencies – starting from 500 MHz to 100 GHz. The BUET study [6], identified the following issues as the reason for wireless service in the buildings becoming increasingly challenging:

- the growth in individual data rate demand,
- the growth in the number of wireless devices, especially, BIoT devices,
- the use of BIoT devices at deeper locations in buildings, and
- the use of higher frequencies.

Ensuring proper radio coverage gets much more challenging, as the frequency increases, which is displayed in the three categories of use cases fit into the spectrum of 5G (**Figure 3**). Evidently, IoT based applications, typically, require low data rate and extended coverage, and thus, they suit lower frequencies. On the other hand, HD videos, VR, and other high data rate applications, require wide bandwidth and thus, high frequencies are used to facilitate eMBB. Similarly, URLLC applications fit in what ranges up to moderately high frequencies.

Building design, clearly plays an important role in wireless connectivity. Undoubtedly, existing buildings will have their own difficulties in addressing the connectivity issues, but proper attention to the related problems is of paramount

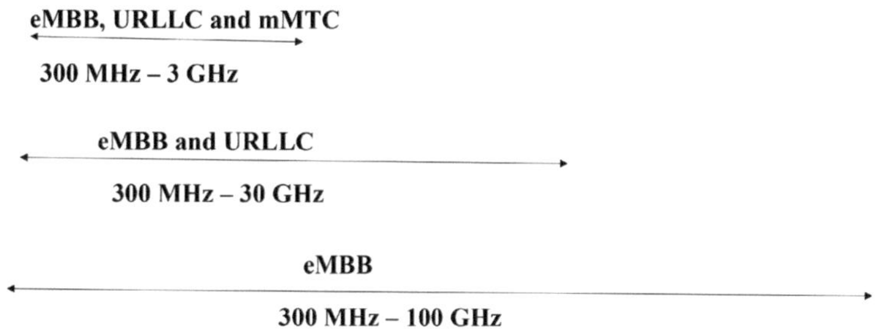

Figure 3.
Distribution of the categories of use cases within the Spectrum for 5G [9].

importance, for making new buildings suitable to this need, at the earliest design stages. This will ensure proper connectivity, as well as, user satisfaction, while addressing any adverse effects that this very propagation may have, on human well-being and health.

6.3 The building interior and internet connectivity

This section looks at internet connectivity, as it is affected by the design of interior spaces. Previous studies, have pointed out the role of penetration losses of various building materials, a quality that largely affects internet connectivity. Good connectivity can cause more homogenous data rate within buildings, thus operating at lower transmit power, conserving their batteries and affecting energy efficiency. An architect's consideration of wireless coverage at the design stage, can help improve coverage significantly. It is an established fact that signal coverage significantly depends on the nature of the space, and its bounding surfaces [3]. However, there is very little work done on establishing these qualities of building materials used in everyday construction.

Addressing this gap, penetration loss levels of some common building materials were determined, as part of the BUET study. The research also related the measured data with other existing information. Based on those measurements, taking into account the scope of architectural design, some guidelines were proposed for architectural intervention, to address the growing challenge of supporting wireless services in buildings. As an outcome of this research, a MATLAB program was developed, using radio propagation theories, which an architect can use during the design phase, to predetermine the impact of the proposed use of different penetration losses of building materials at various frequencies.

6.4 Effects on building design

Studies show that there is a sharp penetration loss at higher frequencies, in typical commercial buildings [10], which use infra-red reflective glass facades, in order to achieve energy efficiency. This will likely have grave consequences on internet connectivity, when the transmission source is outdoors. Propagation losses in interiors, either due to partitions, or space layout, are also considerable at these higher frequencies, and are dependent on the materials used in the layering of the spaces. Such consequences are likely to affect the 'smart' indicators within building interiors, which largely depend on M2M.

Surprisingly, the higher the operating frequency, the faster is the deterioration of radio frequencies, and so the distance, between the transmitting source and receiver in building interiors, needs to be controlled. Another important factor is the path that the wave has to travel between these two points. The higher the frequency, the less its ability to bend around obstacles, therefore requiring more direct visibility/paths between the points. This puts additional restrictions on the design of spaces, than previously encountered. Thus, clearly, both the building structure, and interior partitions can severely obstruct signal strength and internet connectivity, which will result in high propagation losses. This in turn will affect the battery life of the devices, which in itself is challenging under present options, as in many instances they are irreplaceable. Corrective measures like setting up an IBS (in-building solution) is often not feasible for small buildings or residences.

It is important for Architects, as well as interior designers, to be involved in the design and setting up of the wireless connection system, as space layout and the materials chosen, are all decisions taken by the Architect, and an understanding of these issues needs to be one of the considerations, that determine the ultimate design of the interior.

7. Suggested guidelines

This section summarizes the main guidelines suggested as a result of the BUET study [6]. The first of the guidelines concerned the choice of materials for internal partitions. Concrete and infrared reflective (IRR) glass exhibit high penetration losses. Loss due to concrete, takes place on account of it being a very heavy and dense material. Loss from IRR glass, which is not a heavy material, happens due to the reflection of a major part of the signal. On the other hand, plain glass and particle board exhibit low penetration losses, as they are light materials. Also, the higher the number of layers of a material used in a partition, and hence the thickness of the tested material, the higher was the measured penetration loss, with the loss increase being non-linear. In general, clear glass and particle board were found to be low penetration loss materials, while concrete and IRR glass was found to present high penetration losses at the frequencies they were tested for. For higher frequencies the loss would be likely to increase exponentially, pointing to the problems that would be encountered, in a shift from 4G to 5G transmission scenarios.

A stepped process of design was suggested for design to incorporate internet connectivity within buildings. Firstly, Selection of Options for Wireless Service, needs to be considered during building design, suitable for the particular wireless service option chosen. If the wireless signal from an outside cellular base station, seems sufficient for the wireless service in the building, then neither IBS nor Wi-Fi, Zigbee, WiGig, etc. are required. Then the architect should design, ensuring that the signal from outside can enter the building adequately, i.e. taking particular care of the building fenestration.

However, if an IBS is selected, the architect should design for better coverage from the IBS. If Wi-Fi, Zigbee, WiGig, etc. are selected, the installation locations of the sources may be pre-designed in the building, similar to designs produced for electric lights and plumbing. During the design phase, the architect may use indoor radio planning tools, and perform simulations, to check the potential wireless coverage, thereby making valuable adjustments in the architectural design of the building, to improve coverage and signal paths. A few indoor radio planning tools are currently available, like iBwave.

The second step would be to focus on connectivity issues related to architectural design. For any wireless service option, open planning inside the building can help signals propagate better, and pervade throughout the whole building, as the wireless coverage will be dependent on uninterrupted paths within the building. The following points were highlighted to ensure smooth paths and transmission:

i. Open plan can be used, especially, allowing more LOS links for the propagation of radio wave. The open plan concept for buildings is already popular, with less walls to cut off the area, an open plan gives the appearance of more space, which is further compounded by the abundance of light. Since in the case of open plan, the design attempts to avoid the use of real partitions, it allows the best propagation of the radio signal.

ii. Voids, corridors, room size, and so forth, can be incorporated to allow the geometry of space to enhance the openness.

iii. Larger room sizes can be designed for the given design brief.

iv. Long, unobstructed corridors will also allow smooth propagation

v. To improve the link between floors, vertical atriums or voids can be used

vi. Both hard and soft partitions can be used with careful design. An appropriate setting of hard and soft partitions can help allow the desired signal, and block any interference signals. However, this requires proper knowledge and record, of the penetration loss for various building materials, and its variation with increasing frequencies.

vii. The use of glass walls, within open plans can enhance desired signal power. Once the signal strengths are determined, low height walls or partitions can be used towards the desired signal, and high walls or partitions can be used towards any interference signal.

viii. Similarly, thin walls or partitions can be used towards the desired signal and thick walls or partitions can be used towards any interference signal.

ix. The position and orientation of hard or soft partitions can be carefully chosen, depending on the position of user devices, the desired signal source, and the interference signal source.

x. Multiple reflective walls, fringes or louvers may be used carefully to cause multipath signal bounce and get to the receiver.

The suggested guidelines have been presented for the consideration of an architect, but they also create awareness within other professionals, particularly RF Engineers, of the need to collaborate during the design phase, in order to bring relevant connectivity issues to the design board. The possible outcome of such collaboration and the architect's contribution can be summarized as follows.

a. It can be much easier to improve the wireless coverage at the design phase, while the RF engineers will require less time, effort and cost in their installation process.

b. The wireless resources, which are scarce, can be used more efficiently, thus contributing to energy efficiency.

c. The user data rate can be much higher, improving user satisfaction, as well as productivity.

d. IoT devices at deeper locations in buildings can be operable.

e. IoT devices will require less transmit power in uplink and thus, save battery power and meet the requirement of a very long lifetime.

f. The undesired spillage of signal outside the building can be reduced.

8. Covid pandemic issues and other conflicting needs

The recent global pandemic of Covid-19 has also brought focused attention towards sick building syndromes, or SBS. This phenomenon has been a concern for the past five decades, ever since the widespread acceptance of fully air-conditioned buildings became the preferred typology of built spaces, particularly in the thermally challenged situations found in the tropics. The Covid pandemic resulted in the need to maintain social distancing, and in trying to increase the rate at which interior, potentially infected air, is replaced by purer and infection-free outdoor air. Both these requirements have necessitated a shift in the ways in which interiors are conceived.

As the plan layout of spaces is a vital element in maintaining internet connectivity between the transmitting source and receivers, which may be fixed or moveable, these new considerations will also impact the quality of internet connectivity, and needs to be given due thought hand in hand, in order to ensure human health requirements. When more compartmentalization is the need, for isolating infections, and protecting the occupants, the positioning of partitions, their materials and design, all impact the efficiency of internet connectivity.

Green and sustainable planners also advocate compactness in planning a new development, in order to reduce traffic loads, which can be a valuable energy saver. Compactness also allows increased pedestrian movement and biking between destinations, again an active energy efficient measure, which also promotes health benefits from exercise, an added sustainability feature. Again, this measure may contradict the need to create greater distancing between occupants, a requirement vital to control pandemic spreads. The density of neighborhoods is also likely to affect the internet connectivity issue, creating greater obstructions within smaller pathways, affecting the strength of the signals.

The strongly synergistic connection, between the effects of each and every consideration on suitability, regarding physical distancing and/or compartmentalization, is a phenomenon that is encountered time and again, whenever any requirement is compared to others. For instance, the need for avoiding solar exposure may result in infra-red reflective glass facades, but this conflicts with the need to have uninterrupted internet receptivity within the interiors, as mentioned above.

Much research is now required to address the conflicts between the different needs that a building is designed to serve, whether they be thermal comfort, visual comfort, privacy, security, health and air quality needs, and even inclusivity. It is now becoming vital for designers to address the various requirements, and make intelligent and considered choices regarding each, understanding what and the extent of compromises being made for different design decisions, and whether they are potentially harmful or not. The issue of the health hazards of RF transmissions is also of paramount concern at the moment and needs extensive research.

9. Conclusion

The wireless connectivity, in a building, is an important aspect of today's lifestyles, without which it is impossible to function and achieve sustainability. This is because it improves the life of building users greatly, while only consuming nominal energy, making it a key ingredient of green architecture. Thus, it is essentially providing great services, without destroying fossil fuels, and protecting the future world. And this is being demonstrated increasingly, given the work from home scenario found recently during the Covid pandemic. It is unclear how well the World could have handled the lock-down situation, had internet not reached its present development. This makes it of vital importance in the present World, to ensure the provision of seamless internet connectivity, for even the basics of life to function efficiently.

The discussion has related the objectives of the UN SDGs to the issue of having internet access and connectivity, and their intrinsic link to the architecture of spaces. It is difficult in the present times to think of sustainability in the absence of seamless internet connectivity within building interiors. This combines the expertise of multi-disciplinary teams of Architects and RF Engineers.

From a recent research conducted at BUET relating these disciplines, the various options for wireless services have been listed, particularly since the services are increasing their data rate to 5G levels in the near future. Clearly building interiors need to be designed with focus on the issues of seamless propagation of RF waves. The different architectural measures that can be adopted to make this possible have been mentioned here. The geometry of spaces, their spatial flow and materials, their partitions, and openings, all contribute to the flow of internet connectivity. What remains still unaddressed is the matter of the health and safety issues related to 5G transmission scenarios, which it has been suggested deserves special attention in future research.

This paper has brought out the importance of the different disciplines to collaborate in the design of the environment, in order to ensure seamless and safe transmission of internet connectivity. The collaboration needs to begin at the design phase, so that proper decisions are implemented with an understanding of the consequences holistically. Each of the professionals are experts in their own spheres, but they need to make each other aware of the needs which will best serve the built environment, and help improve user satisfaction, while reducing energy wastage. The participatory approach is the only acceptable way forward.

Author details

Zebun Nasreen Ahmed* and Mohammad Tanvir Kawser
Department of Architecture, Bangladesh University of Engineering and Technology, Dhaka, Bangladesh

*Address all correspondence to: znahmed@arch.buet.ac.bd

References

[1] World Economic Forum (WEF); (2015); Fourth Industrial Revolution | World Economic Forum (weforum.org) https://www.usatoday.com/story/sponsor-story/velux/2018/05/15/indoor-generation-and-health-risks-spending-more-time-inside/610289002/ (Accessed December 2020).

[2] Walden, S; (2018); "The "Indoor Generation" and the health risks of spending more time inside What primarily "indoor living" is doing to our health, wellbeing and productivity," Available: https://www.usatoday.com/story/sponsor-story/velux/2018/05/15/indoor-generation-and-health-risks-spending-more-time-inside/610289002/ (Accessed December 2020)

[3] Rappaport, T. (2001); "Wireless Communications: Principles and Practice," Prentice Hall, 2nd Edition, ISBN 0130422320, 2001

[4] GSA Secretariat, KT Ltd., HUAWEI Technologies Co., Ltd., "Indoor 5G Scenario Oriented", White Paper, October 2019.

[5] UNDP; S https://www.undp.org/content/undp/en/home/sustainable-development-goals.html (Accessed December 2020).

[6] Kawser, M. T. (2018, March). Determination of penetration losses of various building materials to help design a building with better wireless services [M. Arch. Thesis; supervisor: ZN Ahmed]. Department of Architecture, Bangladesh University of Engineering and Technology.

[7] Clement, J (2019); "Global Mobile Data Traffic 2017-2022," Statista, Available: https://www.statista.com/statistics/271405/global-mobile-data-traffic-forecast/ (Accessed December 2020).

[8] Waldman, N; (2019); "Connectivity is the next challenge for architects," Available: https://360.here.com/smart-urban-planning (Accessed December 2020).

[9] Johnson, C.; "5G New Radio in Bullets", ISBN: 1077484356, 2019.

[10] E. Semaan, F. Harrysson, A. Furuskär and H. Asplund, "Outdoor-to-Indoor Coverage in High Frequency Bands," 2014 IEEE Globecom Workshops (GC Wkshps), Austin, TX, 2014, pp. 393-398. doi: 10.1109/GLOCOMW.2014.706346

Section 3

Green Technologies and Devices

Chapter 4

Polymer Optical Fiber Splitter Using Tapered Techniques for Green Technology

Latifah Sarah Supian, Mohd Syuhaimi Ab-Rahman, Norhana Arsad, Hadi Guna, Khadijah Ismail, Nik Ghazali Nik Daud, Nani Fadzlina Naim and Harry Ramza

Abstract

Polymer Optical Fiber is opted as the most suitable medium for short haul communication system since it has lower cost and low loss for limited distance of transmission compared to glass fiber. This topic aims to show an alternative, green-technology based, economic and user-oriented communication passive device specifically a directional coupler by lapping tapered-fibers technique. This developed device is using designed geometrical blocks with integration of tapering effect, D_c, macro-bending, R_c, force exertion unto the coupling region, F_c, and etching lengths of the cores, L_e to gain different splitting ratios, i.e., 50:50 and 90:10 experimentally by using the designed geometrical blocks with varied bending radii that affects the radiation of evanescent wave and to relate the integration of Couple Mode Theory and Hertz's Law to obtain optimum coupling efficiency. The development may be an option to current device that are less user-friendly and fragile. This device is developed as a green technology-based device as an option for higher speed communication devices since the materials using in the development is safe, harmless, and inexpensive.

Keywords: polymer optical fiber, acrylic blocks, directional coupler, splitter, geometrical blocks, low-cost, short-haul communication

1. Introduction

This work is conducted to develop an optical fiber passive device based on polymer optical fiber, specifically a splitter or also can be known as a coupler. This device is developed as an effective green-technology based device yet providing an economic solution for home-networking fiber to the home system. The splitter is mainly developed for short-haul communication system where the splitter is developed using polymer optical fiber that has been tapered using harmless organic compound chemical solvent which is acetone. Other method of tapering used in this research is by using side polish where only one side of the diameter of the fiber strands is being tapered. The platform of the device is built using acrylic material having customization of dimensions of the prototype design. Geometrical shapes of circular blocks and ellipse blocks were developed with various bending radii where the tapered fibers are attached to the groove of the blocks and brought closed together. Parameters that involve in this research includes bending radius, R_c, diameter of fiber cores, D_c, coupling length, L_c, and force exertion, F_c.

The coupler/splitter developed in this research includes the preparation of the fibers by etching, side polishing, and building the splitter platform which includes the geometrical blocks with various radii of the macro-bending. The purpose of etching is to eliminate the cladding layers in order to allow the propagation of the modes to travel into the other fibers. However, other factor that could help the transfer of modes from one fiber to the other is macro-bending of the fiber. Evanescent field allows the modes in the fiber to propagate in the cladding. When the fiber bends, losses happen due to the evanescent field that would have to travel faster in order to keep up with the core field. At certain bending, i.e., beyond critical bending, the modes tend to radiate away. In consequence of tapering the multimode optical fiber cladding, higher modes of the fiber are removed while some modes are redistributed.

Other contributing factors that encourage the transfer is the force exertion unto the blocks and fibers attached to the blocks. Some pressure is exerted upon the fibers in order to eliminate any macro-gap that exists between the two parallel fibers. Therefore, when the two fibers are lapped together, the transfer of modes between the two fibers can prevent any leaks of modes that radiates away due to the evanescent field when bending. Length of the parallel coupling also contributes to the effectiveness of mode coupling since when the coupling length is short, only small number of modes get transferred. Coupling length is varied with several bending radii and diameter of the cores to find the optimum performance parameters based on the characterization.

In applying analytical modelling method to characterize and analyze the device, two important theories are used, that are Coupled Mode Theory and Hertz's Law. A simplified couple mode theory between two parallel, lapping multimode step-index fibers are studied where parameters in control which is radius of contact area or coupling length are induced or related by the amount of force or pressure exerted upon the lapping fibers that are attached to geometrical blocks with various radii. The radiation of the propagation modes is induced by the bending of the fiber accordingly to the radius of the geometrical blocks. Depending on the coupling length of the parallel fibers lapped together, the power transfer between them varies in accordance to the length. However, due to physically lapping fibers without any fusion between them, force or pressure is an important aspect in this study so that to eliminate or at least reduce the number of losses of the power transfer due to macro gap. Two important theories are applied which are Hertz's Law and Couple Mode Theory that relates to analytical study of force exertion and radius of contact area.

Polymer optical fibers (POF) show great advantage compared to glass fibers in short-haul communications links due to its flexibility and less expensive, although they are not used for very long distances because of their relatively high attenuation. These characteristics are an advantage for fiber-to-the-home networking having high speed communication. An example would be Internet access within home or within an office [1].

There are several methods that can be used to develop optical fiber coupler/splitter. However, this work aimed to develop optical splitter/coupler that is green-based, safe to use, low cost, economic, easy to install and has multiple solutions for important performance parameters required by users. The optical fiber pairs and the combination of blocks allow the designed platform to produce several performance parameters with minimum loss. Therefore, the device developed can also be a do-it-yourself device since it is customer friendly. The technique used in the process of development is harmless and requires detailed measurement, thus, producing an effective yet low cost POF splitter/coupler that can be used widely by the users.

2. Polymer optical fiber (POF) as lead medium in short distance transmission

Commercially available POF for data communications are polymethilmethacrylate (PMMA) POF core material as shown in **Figure 1**. For visible light of 650 nm, the IR-absorption is 95.9 dB/km, the Rayleigh scattering is 10.3 dB/km and total loss of 106.2 dB/km with no UV-absorption [2]. PMMA POF used in this study is manufactured by Mitsubishi Rayon (Japan). PMMA is produced from ethylene, hydrocyanic acid, and methyl alcohol. It is resistant to water, lyes, diluted acids, petrol, mineral oil and turpentine oil. PMMA tensile strength is approximately 8 kN/cm^2. The refractive index of the core is 1.492 and the cladding is 1.402. The transition temperature lies between +95 $°C$ and 125 $°C$. At room temperature and 50% humidity, the material can absorb up to 1.5% water that can affect the attenuation. The applications include light transmission for signs, illumination, sensors, couplers, nuclear radiation detectors and medical applications.

Polymer optical fiber was introduced in 1960s after glass optical fiber was introduced shortly as a transmission medium for optical communications. Over the years, the transmission capability of POF is improved from having a large attenuation as large as 300 dB/km to 20 dB/km at 650 nm wavelength [3]. POF technology has advantage characteristics such as low insertion loss, low-cost production, thermal stability, mechanical stability, and mass production reliability [3]. Although POFs have higher loss than silica fibers or glass fibers, POFs are never used in long distance communication systems but are being used in intra office communication systems where one requires only a few hundred meters of the fiber. POFs are providing low-cost solutions to short distance applications such as local area networks (LAN), high speed internet access and in vehicles [4].

POF offer the advantages of being lightweight, flexible and easy to handle. Other advantage includes having large fiber cross-section which makes it easier to positioned fiber end at the transmitter or receiver compared to GOF that needs an expensive precision component to center the fiber. PMMA POF has 1 mm diameter

Figure 1.
PMMA polymer optical fiber with diameter 1 mm.

which makes it easy to handle and flexible compared to GOF where the fiber is quite easy to break. PMMA POF is also easy to cut, grind, polish or melt. It also has high flex resistance where the cost used is low even under intense loading conditions that encountered in mechanical engineering applications. Other than that, the easy connecterization of the end faces can be performed cost effectively even after assembling in the field [5].

In respect of electromagnetic compatibility, electrical isolation, immunity to eavesdropping and risk of explosion in hazardous areas, polymer optical fiber and glass fiber have the advantages compared to copper since the photons as the carrier of information in optical fiber have no electrical charge like the electrons which carry the information in copper conductors. In terms of external and mechanical properties, small bending radius and high flexibility are advantages that make POF an attractive choice compared to GOF. The low weight of optical fibers compared with copper is an advantage in most applications.

3. Existing techniques of POF splitter

Couplers work by combining two or more optical signals and combined them into one signal being modulated and propagates through one single fiber whilst splitter in the other hand, separates the signals at the end of the fiber and send the particular signals to their particular destinations [6].

There are three kinds of optical couplers which are directional, distributive, and wavelength-dependent couplers. The mechanism involves in these couplers can be categorized as diffusion type, area-splitting type, and beam-splitting type. Diffusion couplers involve either evanescent wave coupling or radiative coupling. Two fibers are place in proximity [7] and the length of the parallel lapping cores are measured which is known as coupling length. Once they the gap is reduced, radiation of light or known as evanescent wave coupling will initiate thus power transfer will happen. In radiative coupling, bent fibers are coupled to each other by the radiated field. These works well with multimode fibers. Such example is twisted-pair coupler of fused biconical taper coupler. In the fused section, the fiber cores are still separated from each other but the core modes are converted to cladding modes, therefore, partly coupling optical power from one fiber to the other [8]. Example of distributive couplers would be star coupler and example of beam splitter couplers are monitors coupler. This work mainly focuses on directional coupler that has mechanism of diffusion type. Mode selection in multimode fibers has been done by employing offset-launch techniques and mode scramblers by bending the fiber to leak the high order modes and utilized them [9].

Lapping technique is chosen due to the simplicity of the design. Since the development focus on customized and low-cost device, lapping technique could easily be implemented. Other technique such as fused coupler has widely been used and the technique is hardly modified for new approach and new research contributions. Butt coupling and core-facet coupling technique in the other hand has alignment problems and to obtain optimum output will require high-end tools. Y-coupling in the other hand could only produce one output only although the performance is excellent. Other techniques are also discussed in **Table 1** [3, 5].

The demands of couplers include low loss, easy to handle, reproducible coupling behavior, lower manufacturing costs, small dimensions, having thermal and mechanical stability, having low mode dependence and have good isolation between the inputs [5]. Common designations of couplers include 1 × 2, 2 × 2, 1 × N

Type	Technique of fabrication	Advantages	Disadvantages	Loss
Y-coupler	The output fibers are ground where end faces completely cover each other	Has excellent performance as 50:50 power splitters/ couplers	• Costly grinding • Difficult alignment • Losses are caused by the surface of the coupled-in not being fully utilized	EL: 2.7 dB IL: 5.5 dB CR: 1.08 dB Dir: 16.8 dB
Side polishing	Two POF segment are bonded and polished until the core-cladding interface appears.	• The polished depth can be controlled to achieve desired split ratio. • Jacketed POF can be used. • High precision coupling adjustment is possible		NA
Chemical etching	Chemical solvents i.e. acetone, chloroform, methyl isobutyl ketone are used to taper the fiber		• Jacket must be removed • Not easy to control the solution concentration	NA
Reflective body	Device used to split the light is a cylindrical polymer rod.			IL: 4.3 dB

Table 1.
The types of couplers/splitter, the advantages and the disadvantages.

and N × N coupler. Types of couplers/splitters include butt coupler, core fusion coupler, bend coupler and core facet coupler [5].

For coupling to happen using lapping tapered-fibers technique, the two waveguides must be very close so that there is modal overlap, and the coupling coefficient is not zero. The wave is mainly confined within the core thus it is not possible to have wave coupling between fibers by just putting together two fibers side by side. Therefore, a core in one fiber must be very close to the core of another fiber or the propagating wave must extend far outside the core. One of the simple methods is to melt and fused the fiber together. By fusing and tapering the core together, this causes the dimension of the fiber core to be very small, thus the V number (mode number) is small.

Therefore, the propagating waves in the fiber extend far outside the core and coupling occurs according to coupling theory. By properly controlling the dimension of the fiber in the coupling region, a desired ratio of power coupling can be obtained [10]. If the two fibers are identical in the coupling region, both propagating waves will couple or split the same ratio of power from one fiber to another as shown in **Figure 2**.

Most of the existing 1 × 2 splitters only provide one or two fixed splitting ratios. Lapping technique provides the potential of producing multiple splitting ratios by adjusting the coupling length between the two lapping fibers and bending at certain angle. Due to this flexibility of adjusting the coupling length this work is focused on using lapping technique to develop this splitter. In order to produce multiple splitting ratios by bending and tapering, new platform is required to bend the fibers at certain angle and coupling length so that different coupling or splitting behavior

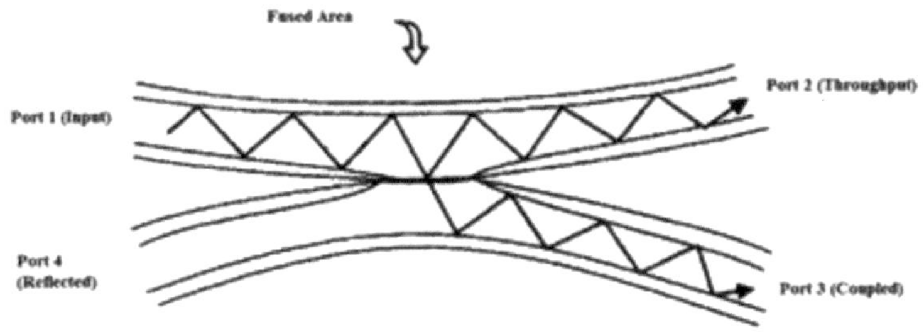

Figure 2.
Lapped fiber coupler with particular lapping length and radiated modes.

or rays will give different splitting ratios. Apart from lapping the fiber at particular length and diameter, certain amount of force is exerted upon the splitter in order to minimize the macro-gap between the fibers. Since parameter of force is also included, study of coupling efficiency between the two lapping curved surfaces with certain amount of load is based on Coupled Mode Theory (CMT) and Hertz's Law. No studies have been done in analyzing the coupling efficiency between the two lapping fibers based on the integration of CMT and Hertz's Law. The coupling efficiency is analyzed when distance, coupling length and fiber diameter is varied.

Ab-Rahman et al. has shown the fabrication of POF coupler/splitter using fusion technique where two POF are melted together and fused to developed N × N coupler/splitter [11]. The modififed coupler/splitter can be extended into demultiplexer. A novel fused POF splitter fabricated by fusion technique is an effective transmission media to split and recombine a number of different wavelengths which represents different signals. The demultiplexer device using different thin film having different colors to filter wavelength and optical splitter that provide optimal results when applied to the data transmission systems [12]. Although fused technique is easy, however, novel approach to develop the splitter is difficult to find. Thus, lapping technique using geometrical blocks are used to develop a directional coupler/splitter.

4. Fabrication using etching technique

There are some methods already done by researchers in order to fabricate coupler/splitters such as fusion between two or more fibers. One of the effective methods is tapering. Tapering can be done for example by technique of stretching fiber whilst it is heated under flame [13], and the other method is by chemical etching. The chemical used, acetone, is safe and harmless and it is effective to remove the cladding layer in certain time. Although tapering may change the physical fiber structure of the fiber itself, however, optical properties mostly remain the same. Due to the core being eliminated, the modes contained in the fiber will be radiated. The radiation of modes may be applied to this study that utilizes couple mode theory.

4.1 Cladding removal by organic solvent

PMMA is dissolved using organic solvents such as acetone and methyl isobutyl ketone (MIBK) in order to remove the polymer in concentric layers as required. Research done by Merchant et al. [13] shows that by using pure acetone without any dilution in water can be used to efficiently remove the cladding layer of PMMA POF. The method requires no tension to be applied on fiber under etching process so as to prevent brittle stress fracture from occurring and break the fiber. The fiber should be supported in a curve and de-stressed fiber is supported in a straight line.

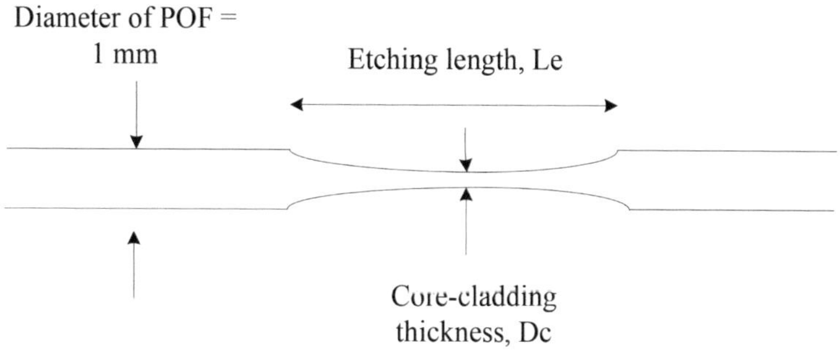

Figure 3.
The waist of the etched region is tapered.

Two to four drops of acetone are applied unto lint-free tissue and it is rotated along the region. The exposed core can be detected as the fiber is decreased in surface friction. Isopropyl alcohol is used to neutralize the solvent and leave the exposed core clean and grease-free. Once the region has been washed, it will return to PMMA physical and chemical properties. Another alternative method is by immersing the fiber into solution containing suitable mixture of organic solvent and water. Even when the solvent is diluted with 20% of distilled water, the brittle property of the fiber during the etching process remain. The fiber region that immerses in the solvent will be uniformly etched producing a linear waist region.

By tapering the multimode optical fiber cladding, higher modes of the fiber are removed while some other modes are redistributed when light source is propagating along the fiber. As the tapered section is developed, the evanescent field and proportion of total power within this field increases in the affected region.

Tapering the fiber can reduce the diameter as shown in **Figure 3** which can filter high-order modes in the fiber and create an effective reduction in numerical aperture which can be an advantage for optical sensor. POF tapers require no alignment and have constant attenuation of low-order modes. The modal redistribution length of POF is a few hundred meters and so the effect of tapers is local to that distance [13].

4.2 Polishing technique

Polishing technique is one of the methods [14–16] to reduce or eliminate the cladding so that the modes that propagate along the fiber may be radiated out due to evanescent wave theorem. Due to polishing effect, which is rough surfaces of the polished fiber, that may lead to increase in losses, therefore, some treatment has to be done. UV curing adhesive having similar refractive index may be used to bridge the gap between the polished fiber in order to reduce the losses. The efficiency of

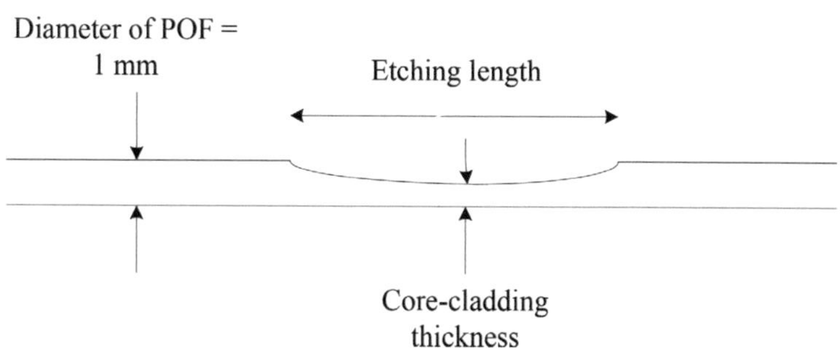

Figure 4.
Side-polished of a fiber.

the coupling light ranging up to 50% and the insertion loss is less than 5 dB [17]. Although it was simple, the limitation occurs when polishing leaves a rugged surface of the fiber. Thus, in this study, the polished surface is done by side etching the surface as shown in **Figure 4** in order to minimize the losses of surface.

5. Macro bending loss by radiation

Losses in optical fiber can be traced back by absorption, scattering or bending. Although losses are not a preferred performance parameter, however, it can be utilized into something useful such as coupling of modes of the bent dielectric waveguide [18]. One of the concepts applies is loss due to macro-bending of tapered bent fiber. The smaller the bending radius, the higher the losses will be [19].

At certain bending radius, which is known as critical radius, the loss is very high where total internal reflection an electromagnetic disturbance which is known as evanescent wave penetrate the reflecting interface. The rate of the propagating evanescent wave will reduce when reflection interface is no longer exists because it cannot propagate in the medium of lower refractive index.

When a light ray hits core-cladding interface, one of the rays will be refracted at the cladding interface and either the ray will be reflected back or refracted with some amount of power [20] while the other ray will propagate at inner core interface. More losses may be observed if bending starts to get smaller in radius where more rays will be refracted so thus the amount of power transferred at the cladding.

Refraction can also cause leaks of rays at the core-cladding interface. Electromagnetic tunnelling at the core-cladding interface is due to the cross section of the curvature. However, the leakage occurring at the tunnelling modes are slower as compared to refracting modes [8]. Some number of rays are not bounded by the core which results in propagating in the cladding region. This is known as the cladding modes and coupling can occur with the higher-order modes of the core resulting in loss of the core power.

There are few benefits of bending losses which are based on either the increase in the attenuation or on making use of the light which escapes from the optic fiber. One of the examples of making use the attenuation experienced by the fiber as it bends is fiber optic pressure sensor where a particular length of bare fiber is placed between two rugged pieces of rubber while the fiber is placed in straight line. A light detector is placed on the end side. When a step pressures the rubber, bends is created, and light intensity is detected, and the alarm went off. On the other hand, an active fiber detector uses light that escapes from the bent fiber. A fiber is placed between jaws of tool and when the fiber is stepped on and pressure exerted upon it, a sharp bend is created by the jaw and some light escapes and detected by the photocell and switch on a warning light [21].

6. Analytical concepts

Simple analytical analysis is studied to analyze the developed coupler using two important concepts which are Coupled Mode Theory and Hertz's Law. Simple coupled mode theory derived by Ogawa [22] analyzes the coupling theory between two parallel multimode step-index fibers and obtaining the coupling efficiency. Hertz's law in the other hand deals with contact mechanics where when load existed between two surfaces that relates to elliptical point contacts and the amount of force on the fibers determines the coupling length of the two fibers.

6.1 Coupled mode theory

Optical directional couplers can be described by coupling length and coupling coefficient as described in Coupled Mode Theory (CMT). The study of CMT has been done among researchers; however, the study of multimode is quite complicated compared to fiber having one or few modes [23]. Thus, this study focuses mainly on two multimode parallel fiber cores using simplified Coupled Mode Theory to find the coupling coefficient and coupling efficiency derived by Ogawa [22]. The coupling efficiency describes the total power of coupling between the two fibers depending on the distance, fiber core thickness and length of the contact region [24].

A multimode coupler or tap coupler is an important component in any short distance communication system. In multimode fiber, it is not easy to evaluate the coupling process between hundreds of modes. Ogawa [22] derived a simplified expression for coupling efficiency between two identical, parallel, step-index multimode fibers which can expand to all modes with a condition that the two fibers are touching each other. Ogawa [22] agrees that the distance between the two fibers affects the coupling efficiency among other considered parameters.

The higher the modes launched at the input of the fiber, the higher the coupling efficiency will be. Higher order modes leakage may result in higher coupling in short lengths [25].

The simplified coupling coefficient given by Ogawa [22] describes that when distance over both radii of core or distance, d is given by D_c, the coupling coefficient becomes as in Eq. (1):

$$C_{coeff} = \frac{2^{1/4} \cdot (n_{co} - n_{cl})^{1/4}}{\sqrt{\pi \cdot k \cdot n_{co}} \cdot a^{3/2}} \left(\frac{i}{N}\right)\left(1 - \frac{i}{N}\right)^{1/4} \qquad (1)$$

Where a = radius of core

$k = \frac{2\pi}{\lambda}.$

n_{co} = refractive index of core

n_{cl} = refractive index of cladding

d = distance between the two fibers

$i = \frac{4}{5} \cdot N$

N = number of modes in step-index multimode fiber

Coupling coefficient reaches maximum when i = 4/5 (N) based on the field interaction between evanescent field of first fiber and second fiber. Higher order modes have stronger field in cladding relative to the field in the core. Ogawa works shows that coupling occurs only between the higher-order modes as the gap increases.

6.2 Hertz's law of elliptical point contacts

Elliptical contact area forms when two 3-dimensional bodies, each with orthogonal radii of curvature come into contact [26]. When force, F is applied between two curved surfaces, compression happens at the beginning of the contact and theoretically a flat surface is formed between them. The area is tangential to the surfaces of the two contacts and it is perpendicular to the line of action of load, F [27].

The radius of the contact area is given by Eq. (2):

$$a = \sqrt[3]{\frac{3 \cdot F \cdot \left(\frac{1 - v_1^2}{E_1} + \frac{1 - v_2^2}{E_2} \right)}{4 \cdot \left(\frac{1}{R_1} + \frac{1}{R_2} \right)}} \quad (2)$$

where E_1 and E_2 are the moduli of elasticity for contact 1 and 2 and v_1 and v_2 are the Poisson's ratios.

The depth of indentation 'd' is related to the maximum contact pressure by Eq. (3):

$$d = \frac{a^2}{R} = \sqrt[3]{\frac{9 \cdot F^2}{16 \cdot R \cdot E^2}} \quad (3)$$

where R is the effective radius defined as shown in Eq. (4):

$$\frac{1}{R} = \frac{1}{R_1} + \frac{1}{R_2} \quad (4)$$

where R_1 and R_2 are the radius of each body respectively.

The size of the circular contact increases weakly with increasing load P and relative radius but decreases weakly with increasing contact modulus. The maximum pressure is 1.5 time the mean pressure and occurs at the center of the contact area [27].

7. New approach using lapping technique

Directional coupler is a passive device where modes exchange between two waveguides that is placed closed to each other. Due to radiation and phenomenon of evanescent wave, some of the power will be transferred to an adjacent guide due to coupling. The factors that contributed to the power or modes exchange between the two parallel fibers are the force exertion and the length of lapping fibers. When two guides are parallel to each other, coupling coefficient is constant and the power launched into one guide will alternate back and forth between the two guides as long as they are close [28].

Lapping technique is the method used in the study where two tapered fibers with certain thicknesses, D_c are placed in proximity at certain length, L_e to each other. Due to the elimination of cladding around the core area, the effective refractive index of the waveguide is varied and coupling coefficient also change [16].

Acetone is a harmless chemical solvent that is used to etch or taper the cladding layer at certain thickness or diameter. The duration of the etching process took around 30 minutes to 120 minutes. Depending on the time of the etching process, the diameter of the tapered fiber will vary. If the cladding layer is decreased, the transfer or power between the lapping fiber will occur. In some cases, the tapering not only causes the cladding layer to be stripped off entirely, but also affect the region where the lapping does not take place which in the other hand resulting to extra losses of the coupler/splitter. Therefore, a platform is developed where mechanical blocks with certain radii will be used together with the tapered fibers having similar refractive index of the fiber that will replace the refractive index of the etched cladding layer.

The varied bending radii, R_c will result to different macro-bending effect, thus, the power transfer between the fiber will also vary accordingly. Coupling length and pressure applied unto the region of the lapping tapered fibers are important in order to obtain high splitting ratio and coupling efficiency.

7.1 Design development

The preparation of the fibers includes preparing the fiber strands of 300 mm long, etching process and side polishing process. Basically, each of the fiber strands was prepared using Mitsubishi Eska Polymethil Methacrylic (PMMA) step-index polymer optical fiber.

The process of etching process is done on polymer optical fiber which has diameter core of ø = 980 μm and diameter cladding of ø = 20 μm thick as shown in **Figure 5**. Chemical solvent that is acetone is used in this study in order to remove cladding layer. Etching process as shown in the figure takes between 30 minutes to 120 minutes to stripped off the cladding layer as intended. Due to the effectiveness of the solvent to impair and remove the cladding layer as reported by the research done by Merchant et al., [29], pure 100% acetone is used in the experiment without any additional liquid or solvent involved or modification the concentration of the solvent.

Figure 6 shows the light transmitting over an etched area between the two blue marks shows faded red light along the area. This fiber has been properly etched and contains the transmitting light with low leakage. Some of the modes are radiated out due to the cladding layers are etched over some duration of time. Therefore, as can be seen in the figure, at the etched region, LED light of wavelength 665 nm are radiated out. The etched fibers are used to develop directional couplers by using geometrical blocks. Etched fibers or denoted as coupler A is fixed unto the circular blocks and they are lapped together. Wavelength of red LED, 665 nm with input power of 16.0 μW is used as the light source. In **Figure 7**, a schematic of bent fibers lapping at certain length with tapered cladding can be seen. Light source having 665 nm wavelength is used to send signal in port a, and the power output are measured at the end of port b, c, and d. The measurements are taken at the end of the

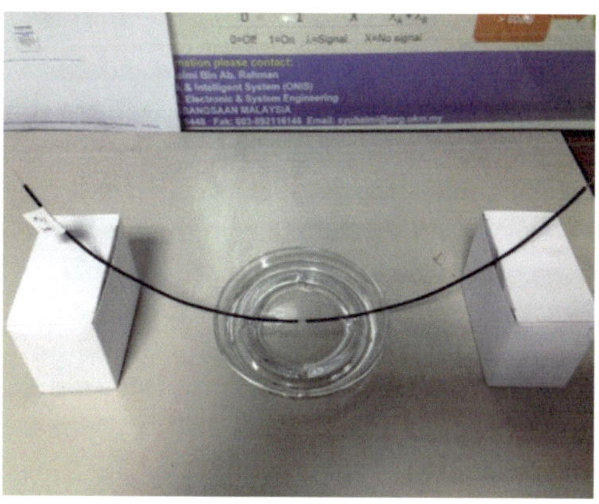

Figure 5.
Etching done by stress-free bending.

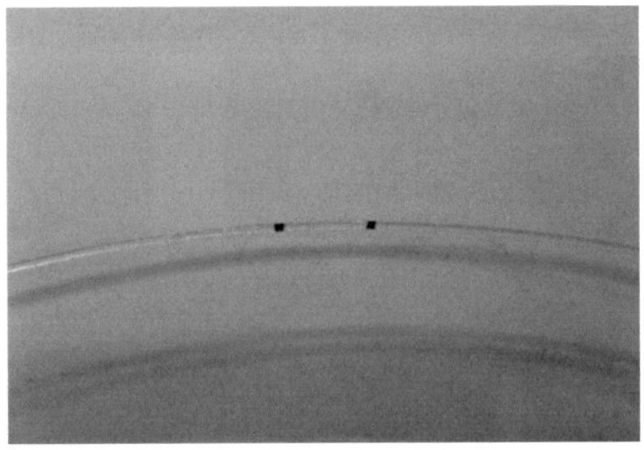

Figure 6.
Light transmission in properly etched fiber.

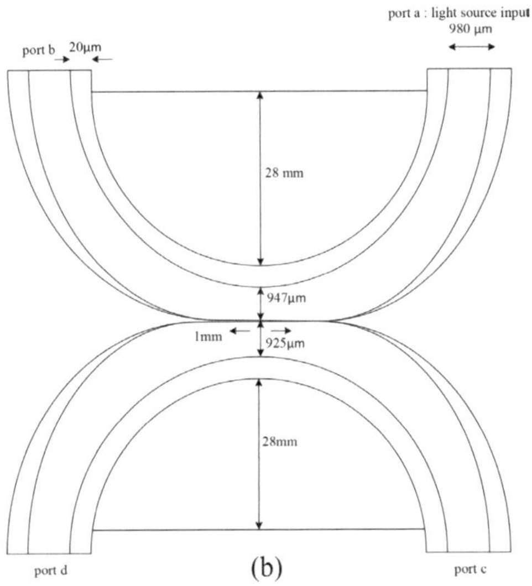

Figure 7.
Schematic of coupler using fibers that were etched.

three output ports in order to analyze the losses due to tapered cladding and force exertion between the two lapping fibers.

7.2 Directional coupler/splitter

There are many efforts done by researchers on developing an optical directional coupler using various techniques as discussed previously. In this work, new technique of developing 1 × 2 optical coupler is fabricated using mechanical techniques where geometrical blocks namely circular blocks of several radii, elliptical blocks of several radii with external forces exerted upon the blocks and fibers and semi-elliptical blocks with spring embedded are used where a pair of etched fibers is placed between them and bent as according to the bending radius of the blocks. Then at input port, P_1, a red LED signal of wavelength 650 nm is injected through the input arm and power meter is placed at the throughput port, P_2, coupled port, P_3 and reflected port, P_4 where the signal strength or output power is obtained and recorded for each output arms. The data recorded are then characterized and analyzed for each pair of coupler/splitter using each pair of blocks with particular bending radius. Characterization such as splitting ratio, insertion loss and excess loss are plotted for each coupler/splitter.

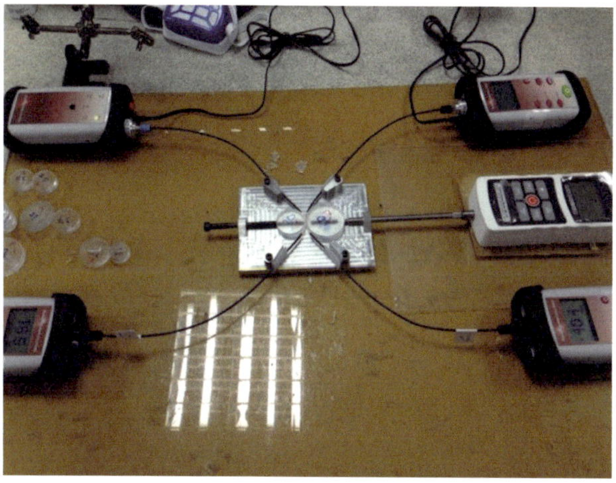

Figure 8.
Circular blocks platform with a pair of tapered fibers bent according to the bending radius of the blocks and input of 650 nm is inserted into one of the input and the output power is recorded.

Figure 8 shows the setup of circular blocks where the fibers length are around 30 cm using Mitsubishi Rayon Eska POF. The end of each fiber is connected with power meter of type AF OM 210A.

Table 2 shows the optimum splitter at each circular blocks pair. Splitter of coupling diameter, D_c of 0.92 mm–0.90 mm and 0.85 mm–0.75 mm give the optimum value of splitting ratios, SR for most of the circular blocks. Insertion loss, IL and excess loss, EL can be referred in the table accordingly. The reason might be due to the thickness of the core-cladding layers of each splitter that allow some of the light rays to propagate along the tapered length and when the fiber is bent accordingly and lapped to the tapered region of second fiber, the rays is coupled in the second fiber. Due to the fully etched region around the fibers that is not lapped to other waveguide and the extra tapered length, high losses are observed.

Although splitter of D_c of 0.92 mm–0.90 mm shows the most optimum SR_c for circular blocks, however, the excess losses are also the highest followed by D_c of 0.92 mm- 0.94 mm and the lowest excess loss (EL) is given by splitter of D_c of 0.95 mm- 0.95 mm as in **Figure 9**. Since most of the rays were transferred by the

Rc (mm)	Dc (mm)	SRc (%)	EL (dB)	IL (dB)
25–25	0.92–0.90	2.41	6.00	22.00
30–40	0.85–0.75	2.00	3.00	20.00
35–27	0.85–0.75	2.30	3.00	19.60
30–20	0.92–0.90	2.60	6.00	22.00
52–40	0.85–0.75	1.00	3.50	23.00
28–26	0.85–0.90	1.30	2.70	21.00
28–22	0.85–0.75	2.50	3.50	19.60
28–23	0.85–0.75	1.70	3.80	21.00
38–34	0.85–0.90	1.00	2.80	22.00
38–37	0.85–0.75	1.80	3.80	21.00

Table 2.
Splitters with optimum SR, EL and IL for each bending radius.

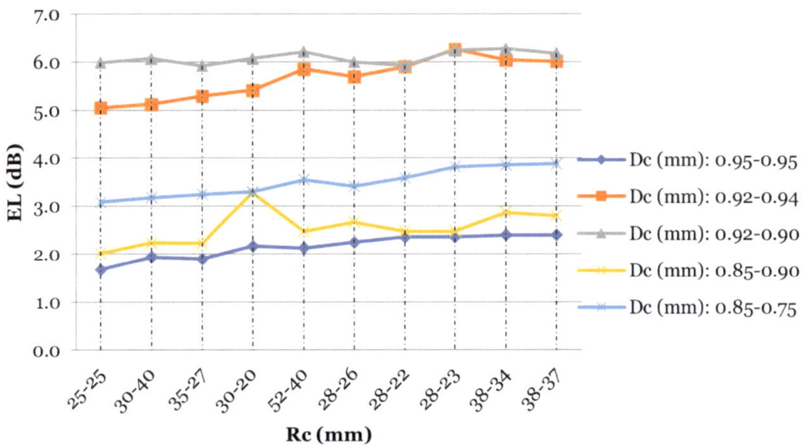

Figure 9.
Excess losses for each coupler/splitter of different core-cladding Thickness, D_c, for each circular blocks of different bending radii, R_c.

first fiber to the second fiber for D_c of 0.92 mm –0.90 mm, due to the unlapped region of the tapered section, many of the rays radiated out by small bending of circular blocks and the small coupling length between the two lapped regions contributed to the losses. However, for splitter of D_c of 0.95 mm- 0.95 mm, since the cladding layers conserved many of the rays from being radiated out of the fibers, the excess losses are lower for all the bending radii of the circular blocks.

At throughput port as shown in **Figure 10**, splitters of D_c of 0.92 mm- 0.94 mm and D_c of 0.92 mm –0.90 mm show the highest losses due to the radiation of the first fiber to the second fiber whilst D_c of 0.95 mm- 0.95 mm shows the lowest insertion loss at throughput port since the cladding layers that existed in the fibers prevent the rays from being transferred or radiated out from the first fiber.

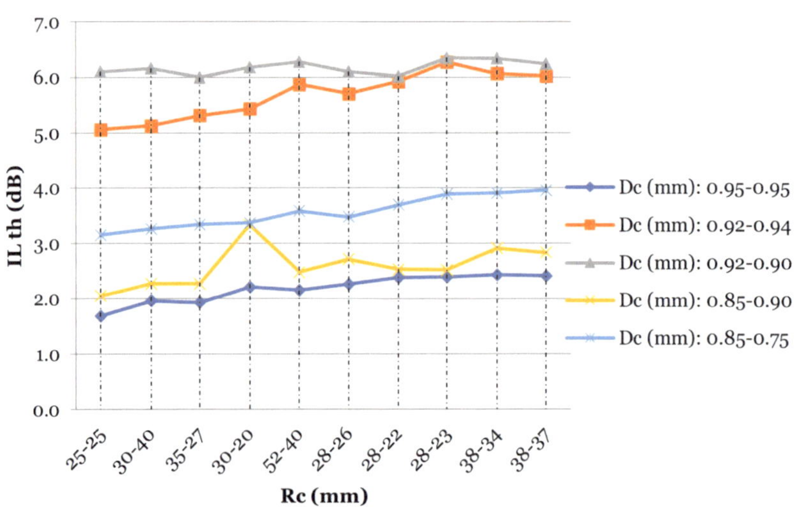

Figure 10.
Insertion losses at throughput port for each coupler/splitter of different core-cladding thickness, D_c, for each circular blocks of different bending radii, R_c.

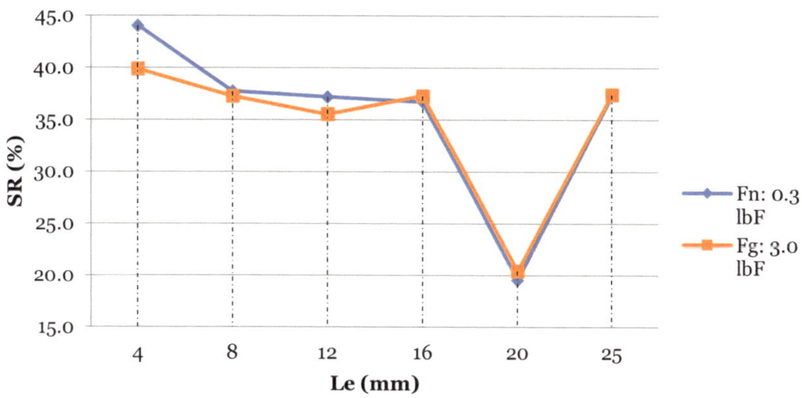

Figure 11.
Average splitting ratio for normal and external load unto splitter of varied etching length fiber splitters using varied combination of circular blocks.

7.3 Load exertion

Figure 11 shows the average splitting ratios when external load is exerted upon the splitters fitted into circular blocks. Splitting ratios show a slight decreased when external load is exerted upon the splitters of etching length, L_e of 4 mm, 8 mm, 12 mm. However, splitting ratios increase when load is exerted upon longer tapered length of L_e of 16 mm, 20 mm and 25 mm. Splitting ratios of L_e of 4 mm decreases from 44% to 40%, for L_e of 8 mm, SR decreases from 38% to 37% and for L_e of 12 mm, SR decreases from 37% to 36%. Splitting ratios for all other splitters increase about 1% when external force is exerted.

8. Comparison of efficiency between experimental and analytical values at particular coupling length

The coupling efficiency from experimental values are calculated by

$$\eta = \frac{P_o}{P_i} \times 100 \qquad (5)$$

The efficiency of coupling length, L_c of 3 mm, 5 mm and 8 mm are 21.69%, 29.3% and 28.7% respectively is compared to analytical values simulated in MathCAD using Eq. (5) as shown in **Figure 12**.

Analytically, the result shows the coupling efficiency of core radius of 0.75 mm at coupling length extends from 0 mm to 20 mm. The distance, d, between the two cores is assumed 1 μm. Coupling efficiency from the graph at specific L_c, namely 3 mm, 5 mm and 8 mm is found to be 29%, 52% and 70% respectively.

It is observed that the pattern of the coupling efficiency of experimental values and analytical values is similar as shown in **Figure 13**, however, analytically the ideal simulated wave shows higher percentage of efficiency compared to the values of the efficiency of the experiment. The average difference of efficiency between analytical and experimental values is between 7% to 22%. The differences between the values are due to several factors. Due to the varied bending angle, R_c, of the fibers attached to the semi-elliptical blocks, the radiation of rays from the input fiber to the second

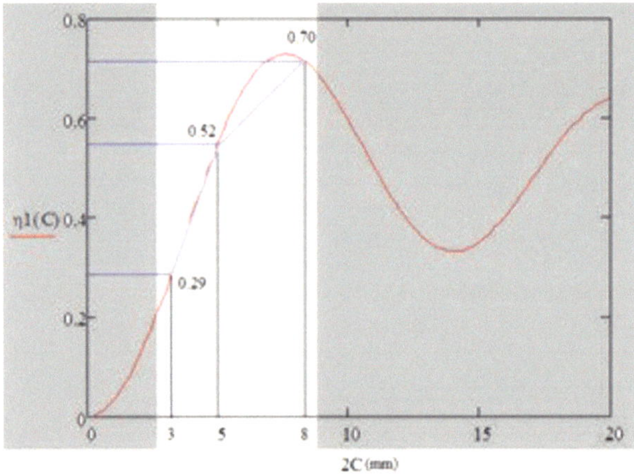

Figure 12.
Efficiency values at coupling Length.

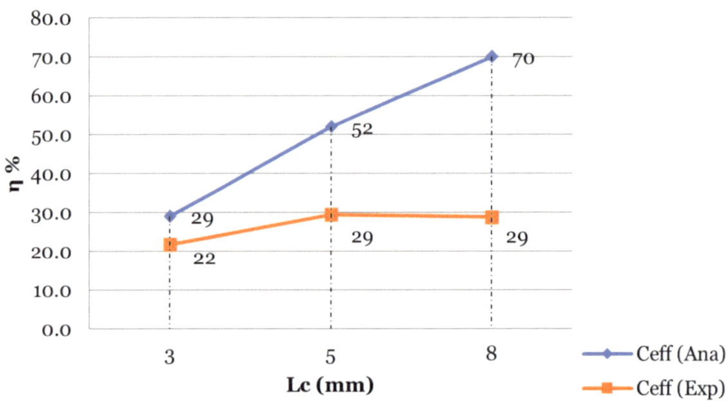

Figure 13.
Similar pattern of efficiency is observed for both analytical and experimental values at specified coupling length.

fiber varies accordingly. At 30 mm bending radius, the smaller bend stimulates the rays to radiate out more extensively, however, the losses are high due to tapered length that is longer than 3 mm which reduces the efficiency. At 40 mm, since the coupling length is longer which is 5 mm, the efficiency is higher. However, at 50 mm bending radius, although the coupling length is longer, however, the bending angle is larger which hinders the stimulation of rays to radiate out that causes the efficiency to reduce again. The etching or tapered length also causes the efficiency of the experiment to reduce since the extra tapered length could contribute the radiation of rays at the non-lapping region. High efficiency is found at simulated graph since the condition of the coupling is ideal where the condition of coupling in experiment accounts several factors that lowers the efficiency such as the tapered length of the cores, bending angle, insertion losses and connection of the fibers to the power meters.

9. Conclusion

New technique of developing an optical coupler using POF and mechanical platform using lapping technique is discussed and analyzed. The device fabricated is flexible since the splitter does not only give one particular splitting ratio desired, but it can be customized using different blocks of bending radius and several pair of fibers with different core-cladding thickness. The different pair of fibers can be matched with different pair of blocks in order to obtain particular splitting ratio for different applications. The implementation is simple where the blocks need to be placed on the platform where spring-like component will force the blocks to hold the fiber pairs close in proximity. Analysis of efficiency between experimental values of splitters and values obtained by simulated analytical values are compared where similar pattern of efficiency behavior is observed for the splitters which shows that the splitter is good to be used.

The effect of different angle of fiber bending integrates with different taper length and core diameter with force exertion is studied and analyzed in this study. Optimum results of splitting ratios is obtained by having bending radius between 30 mm to 50 mm, taper length between 4 mm to 20 mm, core thickness between 0.88 mm to 0.77 mm, coupling length between 4 mm to 10 mm and 18 mm to 22 mm and pressure not less than 3.0 lbF. The variation of the parameters leads to different coupling characteristics thus resulting in various splitting ratios and losses. Therefore, for different parameter values, particular values of other parameters have to be considered.

The maintenance of the device is also simple. The fiber pairs can be used continuously and interchanged with other fiber pairs. The fibers need to be changed with new ones only when the fiber is broken. In this case, since POF itself is very flexible, thus, the flexibility and maintenance of the fibers are quite reliable. The platform and the blocks are made of strong material that is hardly broken even they are dropped several times. Even in high temperature and heat, POF melting point is around 80 $°C$ whilst the blocks that are made of acrylic material have melting point of 160 $°C$. The platform based is made of aluminum that have very high melting point. Thus, the device can be used inside medium heat compartment such as in automobile. The reproducibility of the device is in the other hand in small production due to the hand-made fabrication at the research stage. This device is green technology based since the production process is using eco-friendly material, harmless solvent and the LED source is used which is very safe for consumers as compared to other method that use the heat using burner can contributes to CO_2 even in small amount.

The developed device can be innovated into a 'DIY' kit where the installation of the passive device will be easy and customer friendly. Different values of splitting ratios are able to be achieved thus give the advantage of different applications for the users. Moreover, since the device may provide different values of splitting or coupling ratios in one kit, users may no longer need to spend extra on purchasing another splitter/coupler. This device is inexpensive and green technology based due to materials used in the development and utilization.

Author details

Latifah Sarah Supian[1]*, Mohd Syuhaimi Ab-Rahman[2], Norhana Arsad[2], Hadi Guna[3], Khadijah Ismail[1], Nik Ghazali Nik Daud[1], Nani Fadzlina Naim[4] and Harry Ramza[5]

1 National Defence University of Malaysia, Kuala Lumpur, Malaysia

2 Universiti Kebangsaan Malaysia, Bangi, Malaysia

3 Universiti Tunku Abdul Rahman, Kajang, Malaysia

4 Universiti Teknologi MARA, Shah Alam, Malaysia

5 Universitas Muhammadiyah Prof. Dr. Hamka, Jakarta, Indonesia

*Address all correspondence to: cawa711@gmail.com

References

[1] Haupt, M., Reinboth, C. & Fischer, U.H.P. Realization of an economical polymer optical demultiplexer. International Students and Young Scientists Workshop; 2006.

[2] IGI Consulting. Plastic Optical Fiber Market & Technology Assessment Study. IGI Group; 2011.

[3] Nalwa, H.S. Polymer Optical Fiber. California: American Scientific Publisher; 2004.

[4] Ghatak, A. Optics. New Delhi: McGraw Hill Education; 2013.

[5] Weinert, A. Plastic Optical fibers: Principles, Components, Installation. Publicis MCD Verlag; 1999.

[6] Chen, C.L. Optical directional couplers and their applications. John Wiley & Sons; 2007.

[7] Love, J.D. & Durniak, C. Bend loss, tapering and cladding-mode coupling in single-mode fibers. IEEE Photonics Technology Letters. 2007; 19(16): 1257-1259.

[8] Agarwal, D.C. Fibre Optic Communication. Wheeler Publishing; 1993.

[9] Kim, D.G., Woo, S.Y., Kim, D.K. & Park, S.H. Fabrication and characteristics of plastic optical fiber directional couplers. Journal of the Optical Society of Korea. 2005; 9(3): 99-102.

[10] Lin, C.F. Optical Components for Communications: Principles and Applications. Masschusetts: Kluwer Academic Publishers; 2010.

[11] Ab-Rahman, M.S., Guna, H., Harun, M.H., Supian, L. & Jumari, K. Integration of Eco-friendly splitter and optical filter for low-cost WDM network solution. Optical Fiber Communication and Devices. InTech; 2012.

[12] Harun, M.H., Ab-Rahman, M.S. & Safnal, M.H.G. Improving performance of handwork-fused 1×3 polymer optical fiber splitter through progressed fusion technique. In: International Conference on Computer Technology and Development; 2009.

[13] Merchant, D.F., Scully, P.J. & Schmitt, N.F. Chemical tapering of polymer optical fiber. Sensors and Actuators A: Physical Elsevier. 2000; 76(1-3): 365-371.

[14] Ji, F., Xu, L., Li, F., Gu, C., Gao, K. & Ming, H. Simulation and experimental research on polymer fiber mode selection polished coupler. Chinese Optics Letter. 2008; 6(1).

[15] Tanaka, T., Serizawa, H. & Tsujimoto, Y. Characteristics of directional couplers with lapped multimode fibers. Applied Optics. 1980; 19(20): 2019-2024.

[16] Findakly, T. & Chen, C.L. Optical directional couplers with variable spacing. Applied Optics. 1978; 17(5): 769-773.

[17] Kawase, L.R., Santos, J.C., Silva, L.P.C., Ribeiro, R.M. Canedo, J. & Werneck, M.M. Comparison of different fabrication techniques for POF couplers. In: The International POF Technical Conference, Cambridge, Massachusetts, USA; 2000. 68-71.

[18] Gloge, D. Bending loss in multimode fibers with graded and ungraded core index. Applied Optics. 1972; 11(11): 2506-2513.

[19] Barnoski, M.K. & Friedrich, H.R. Fabrication of an access coupler with single-strand multimode fiber waveguides. Applied Optics. 1976; 15(11): 2629-2630.

[20] Durana, G., Zubia, J., Arrue,J., Aldabaldetreku, G. & Mateo., J. Dependence of bending losses on cladding thickness in plastic optical fibers. Applied Optics. 2003; 42:997-1002.

[21] Crisp, J. Introduction to Fiber Optics. Newnes; 1996.

[22] Ogawa, K. Simplified theory of the multimode fiber coupler. The Bell System Technical Journal. 1977; 56(5):729-745.

[23] Marcuse, D. Field deformation and loss caused by curvature of optical fibers. Journal of Optical Society of America. 1976; 66(4): 311-320.

[24] Haus, H. A. & Huang, W.P. Couple-mode theory. In: Proceedings of the IEEE; 1991. 79(10):1505-1518.

[25] Ogawa, K. & McCormick, A.R. Multimode fiber coupler. Applied Optics. 1978; 17(13): 2077-2079.

[26] Hale, L.C. Principles and techniques for designing precision machines [thesis]. MIT; 1999.

[27] Johnson, K.L. Contact mechanics. Cambridge: Cambridge University Press; 1985.

[28] Badar, A.H., Maclean, T.S.M., Gazey, B.K., Miller, J.F. & Shiraz, H.G. Radiation from circular bends in multimode and single-mode optical fibres. In: IEEE Proceedings; 1989. 136(3): 147-151.

[29] Merchant, D.F., Scully, P.J. & Schmitt, N.F. Chemical tapering of polymer optical fiber. Elsevier; 1998.

Chapter 5

Eco-Friendly, Green Approach for Synthesis of Bio-Active Novel 3-Aminoindazole Derivatives

Chandrashekhar Devkate, Satish Kola, Mohammad Idrees, Naqui J. Siddiqui and Roshan D. Nasare

Abstract

In present chapter we have reported green and highly efficient method for synthesize novel series of substituted -1H-indazol-3-amine derivative (**3a-h**) by cyclocondensation reaction of substituted benzonitrile (**1a-h**) and substituted Hydrazine (**2a-h**) using ceric (IV) ammonium nitrate (CAN) as a catalyst, EtOH-H_2O as a ecofriendly media and reaction was carried out under ultrasound irradiation green method. The structures of newly synthesized indazole derivative (**3a-h**) were corroborated through spectral investigation such as elemental analysis and spectral studies like IR, C^{13} NMR, Mass spectra and ^{1}H NMR. The compounds were assessed for their in-vitro antimicrobial activity with pathogenic microbe comprising Gram positive bacterial strains, *S. aureus* and Gram negative strains *E.coli, P.vulgaris,* and *S. typhi* at different concentration. The consequence of bioassay is compared with standard drug Chloramphenicol.

Keywords: indazol, ceric (IV) ammonium nitrate catalyst, ultrasound irradiation, ecofriendly media, antimicrobial screening

1. Introduction

Indazole was first defined as a "pyrazole ring fused with the benzene ring" by the scientist Emil Fisher. It is broadly studied due to its remarkable chemical and biological properties. Indazole is from the azoles family containing carbon, hydrogen and nitrogen atoms. Indazole also called as benzpyrazole or isoindazolone which containing two nitrogen atoms. It is ten π-electron aromatic heterocyclic systems as a pyrazole molecule. The structure of indazole is given below in cylindrical bonds is as (**Figure 1**).

Indazole derivatives are pharmacologically significant as they form the fundamental structure of numerous drug molecules, like Benzydamine used as anti-inflammatory agent and Granisetron, 5HT3 receptor antagonist for anti-emetic in cancer chemotherapy. Two nitrogen atoms in indazole can be able to be functionalized with high selectivity at different positions. Indazole show a range of biological activity such as anti-HIV, anti-cancer, anti-platelet, anti-inflammatory, serotonin 5-HT3 receptor antagonist and anti-tumouractivities [1–6]. 3-Aminoindazoles which are valuable templates for medicinal chemistry. The scaffold is found in a huge number of compounds exhibiting a large number of biological activities including kinase inhibitors, HIV protease inhibitors, MCH

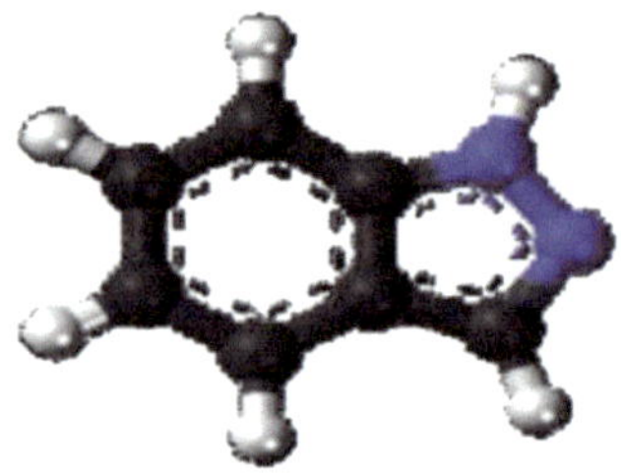

Figure 1.
Naturally accruing indazole nucleus.

receptor1 antagonists, CB1 receptor inhibitors and factor XIa inhibitors [7–9]. The 3-aminoindazoles have been able to mimic the adenine nucleus of ATP for the design of ATP-competitive receptor tyrosine kinase inhibitors with potent antitumor activities. Thus it is been potential valuable templates for pharmaceutical chemistry, hence effort has recently been taken to the synthesis of substituted indazole. Several method have been published for the synthesis of 3-aminoindazole the methods have several drawback is the use of costly reagents and catalyst, organic solvents, harsh conditions and thus have limited scope [10–13]. In last few decades Chemist have attraction for Nitrogen containing heterocycle which possess potential antimicrobial property [14–16]. Indazole can be present in two forms which result from the displacement of proton among two nitrogen atoms, a method describe as proto tropic annular tautomerism. Without substituted indazole be present mostly as the 1H-tautomer based on the results from molecular studies.

The indazole ring contains two nitrogen atoms and thus annular tautomerism with concern to the location of the NH hydrogen atom. The variation in energy among tautomer's the benzenoid type predominates in the gas phase solution. Solid state derivatives are generally found thermodynamically more constant than the analogous 2H forms, annular tautomerism of indazole benzenoid 1H-indazole tautomer and quinonoid 2H-indazole tautomer. Ortho-hydrazine benzoic acid on heating results in the formation of indazolone reported by Emil Fisher in 1800 (**Figure 2**) [17–20].

Biological importance of 3-aminoindazole Derivatives:- Indazoles are naturally occurring alkaloids like Nigellidine, Nigellicine and Nigeglanine. Nigellicinewas isolated from extensively spread plant *Nigella sativa*. Nigeglaninewas isolated from extracts of Nigella glandulifera. Merely few of the alkaloids studied upon isolation show the presence of indazole ring system. The first member of this alkaloid family Nigellicine [21] which is isolated in 1985 from the plant *N. sativa* an annual flowering plant, native to Southwest Asia. The seeds of this plant are used for thousands of years as a spice and for the treatment of a variety of diseases [22–24]. The structure of nigellicine has an intramolecular hydrogen bond among the carboxylate oxygen atom and the hydroxyl group. The structure of nigellicine is a pseudo cross conjugated heterocyclic mesomericbetaine, which means that it be able to be presented by dipolar canonical formula where both the positive and negative charge is delocalized in the structure [25].

Similarly other two alkaloids (**Figure 3**) Nigeglanine and Nigellidine are isolated from extracts of *N. glandulifera* and *N. sativa*. These two compounds be able to also be obtainable by their zwitterions formulae [26]. Indazole core is present in naturally occurring alkaloids and biologically active molecules. Nigellidine is a natural product containing an indazole nucleus, isolated from plant *N. sativa* and used in the treatment of a variety of diseases. Indazole and their derivatives are found to have a large range of activities. Previous findings on indazole derivatives are purposely known to be active as protein kinase inhibitors, cancer cell proliferative

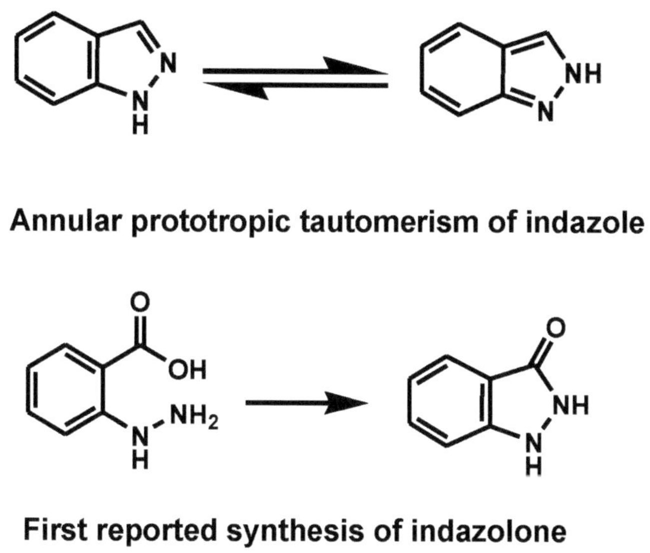

Figure 2.
Naturally accruing indazole nucleus.

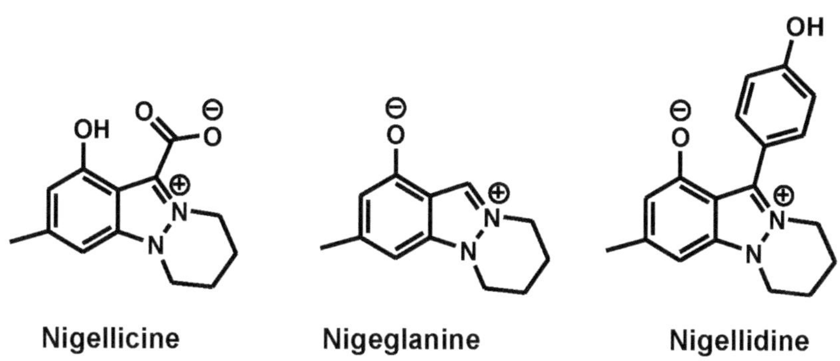

Figure 3.
Naturally accruing indazole nucleus.

disorders, the methods 3-aminoindazole have several drawback is the use of costly reagents and catalyst, organic solvents, harsh conditions and thus have limited scope [27, 28].

Stimulated from these observation and literature exploration revealed that no green and efficient, method is reported yet hence in the present work we are endeavored to synthesize 3-aminoindazole by using ultrasonic radiation and ceric ammonium nitrate (CAN) [Ce IV(NO_3)6]2 as a catalyze reactions in organic synthesis because of many advantages such as good solubility in water, easily available, non-toxicity, simple work-up procedure, low cost, high reactivity and synthesized substituted-1H-indazol-3-amine derivative (3a-h) and screened for their antimicrobial activity [29–34].

2. Material and methods

Chemicals used for the synthesis were of AR grade of Merck, S.D. Fine and Aldrich. The reactions were examined by E. Merck TLC aluminum sheet silica gel60F254 and visualizing the spot in UV Cabinet and iodine chamber. The melting points were note down in open capillary in paraffin bath and are uncorrected. ^1H NMR spectra are logged on a Bruker AM 400 instrument (400 MHz) using tetramethylsilane (TMS) as an internal reference and DMSO-d6 as solvent. Chemical Shifts are specified in parts per million (ppm). Positive-ion Electro Spray Ionization (ESI) mass spectra were acquired with a Waters Micromass Q–TOF Micro, Mass Spectrophotometer. IR spectra were recorded on a Shimadzu IR Spectrophotometer (KBr, υmax in cm^{-1}). The compounds are purified by using column chromatography on silica gel (60–120 mesh). Elemental (CHN) examination was done using Thermo Scientific (Flash-2000), the compounds were investigated for carbon, hydrogen and nitrogen and the results found are in good agreement with the calculated values.

3. Experimental section

3.1 The optimization of the reaction conditions

The optimization of the reaction is done using different solvents and also solvent free condition was applied and the reaction carried out under ultrasound irradiation. When the reaction was performed without solvent the yield obtained was in trace amount. When different solvent like acetonitrile, toluene and ethanol it the yield was less and the time taken for completion of reaction was more. But using (CAN) (10 mol %) EtOH-H_2O (2:2) (entry 9) at 50–60°C for 35 min. we got good yield in less time and thus the reaction was optimized. The results were summarized in (**Table 1**).

General Procedure for the synthesis 3-aminoindazole (**3a-h**):- A mixture of benzonitrile (**1a-h**) (1.0 mmol), hydrazine (**2a-h**) (1.2 mmol) and (CAN) (10 mol) in solvent EtOH-H_2O (2:2) were taken in single neck round bottom flask and the flask containing reaction mixture was kept in the ultrasonic bath and was irradiated at 50-60°C for about 30–40 min. (the progress of reaction was monitored by TLC at different interval) separately as indicated in (**Table 2**). After the reaction was completed the reaction mass was poured on crushed ice. The obtained solid was

Entry	CAN mol (%)	Solvent	Time (min)	Yield (%)
1	—	—	80	5
2	5	MeCN	60	40
3	10	MeCN	60	58
4	5	Toluene	70	53
5	10	Toluene	70	60
6	5	EtOH	60	68
7	10	EtOH	60	76
8	5	EtOH-H_2O (2:2)	35	80
9	10	EtOH-H_2O (2:2)	**35**	**93**
10	15	EtOH-H_2O (2:2)	35	86

Table 1.
Optimizing of the reaction conditions for 5-bromo-1-methyl-1H-indazol-3-amine (3c).

Comp.	R	R'	product	m.p (°C)	Yield (%)
3a	H	Me	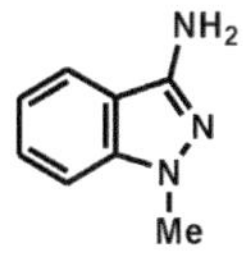	95–97	90
3b	H	H	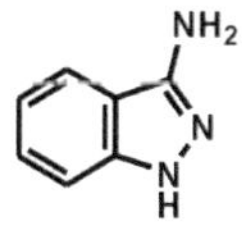	150–152	87
3c	5-Br	Me	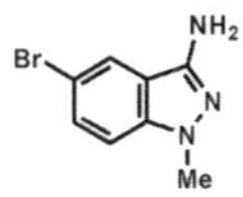	133–135	93
3d	5-Cl	Me	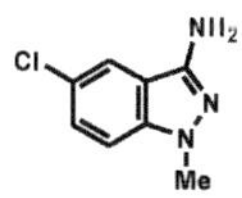	130–132	89
3e	6-OMe	H	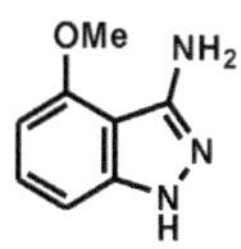	116–120	90
3f	4-Br	H	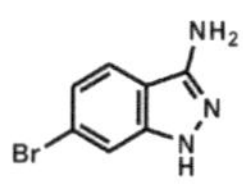	171–174	95
3 g	5-I	H	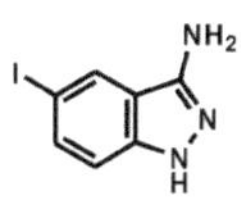	177–179	93
3 h	5-F	H	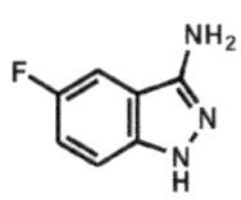	165–167	90

Table 2.
CAN catalyze synthesis of 3-aminoindazole under ultrasound irradiation (3a-3 h).

filtered, washed with water and dried. The crude compound was crystallized using DMF-Ethanol. Their structure was confirmed by ^{1}H NMR, physical data, mass, IR and elemental analysis.

1-methyl-1H-indazol-3-amine (3a):

IR (KBr, υ_{max} in cm^{-1}): 3422 (N-H str., -Amine),2982,2942 (C-H asym. Str. aliphatic), 2830 (C-H sym. Str., aliphatic), 3054 (C-H str., aromatic),1540 (C=N str., Indazolyl), 1342 (C-N str.), 1504,1465 (C=C str., aromatic), 1119, 1126 (C-H i. p. def., aromatic), 867 (C-H o.o.p. def., aromatic).

^{1}H NMR δ ppm (DMSO-d_6): δ 3.79 (s, 3H, -CH_3), 5.36 (s, 2H, Amine), 7.2–7.8 (m,4H, Ar-H).

^{13}C-NMR (100 MHz, DMSO): δ 35.52, 110.06, 112.34, 118.15, 122.83, 128.64, 139.45, 147.50 ppm.

LCMS (ESI+) *m/z*: 148[M + 1]$^{+}$, 271[M + Na]$^{+}$, 249[M + Na + H]$^{+}$.

Elemental Analysis: Calcd. For $C_8H_9N_3$; calculated: C, 65.29; H, 6.16; N, 28.55 Found: C, 64.54; H, 6.19; N, 27.95.

General Reaction:- Synthesis of compounds Indazole (3a-3 h) Derivatives.

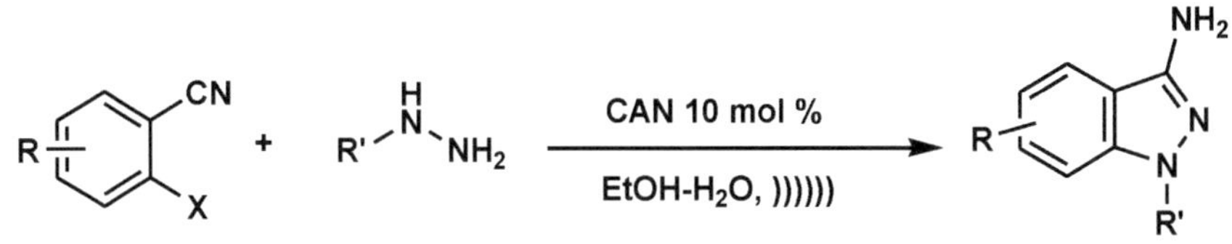

Where, R = H, Br, Cl, OMe, I, F.

R' = Me, H

1-H-Indazol-3-amine (3b):

IR (KBr, υ_{max} in cm^{-1}): 3420 (N-H str., -Amine), 2980, 2941 (C-Hasym. str. aliphatic), 2833 (C-H sym. Str., aliphatic), 3051 (C-H str., aromatic),1544 (C=N str., Indazolyl), 1343 (C-N str.), 1503,1464 (C=C str., aromatic), 1116, 1123 (C-H i. p. def., aromatic), 864 (C-H o.o.p. def., aromatic),554 (C-Br str., Ar-Br).

^{1}H NMR δ ppm (DMSO-d_6): 5.41 (s, 2H, -NH_2), 12.4 (s, 1H, ring-N-H), 7.2–7.7 (m,4H, Ar-H).

^{13}C NMR δ ppm (DMSO-d_6): δ 41.55, 108.01, 113.34, 117.75, 126.88, 128.62, 141.59,

LCMS (ESI+) *m/z*: 134[M + 1]$^+$,157[M + Na]$^+$, 158[M + Na + H]$^+$.

Elemental Analysis: Calcd. For $C_7H_7N_3$; calculated: C, 63.14; H, 5.30; N, 31.56Found: C, 63.17; H, 5.34; N, 31.34.

5-bromo-1-methyl-1H-indazol-3-amine (3c):

IR (KBr, υ_{max} in cm^{-1}): 3424 (N-H str., -Amine), 2983,2943 (C-H asym. Str. aliphatic), 2834 (C-H sym. Str., aliphatic), 3052 (C-H str., aromatic),1541 (C=N str., Indazolyl), 1344 (C-N str.), 1501,1464 (C=C str., aromatic), 1116, 1122 (C-H i. p. def., aromatic), 861 (C-H o.o.p. def., aromatic),556 (C-Br str., Ar-Br).

^{1}H NMR δ ppm (DMSO-d_6): 3.65 (s, 3H, -CH_3), 5.47 (s, 2H, -NH_2), 7.5–7.6 (d, 2H, Ar-H), 7.93 (s, 1H, Ar-H).

^{13}C NMR δ ppm (DMSO-d_6): 34.55, 109.01, 110.34, 115.75, 122.88, 128.62, 139.59, 147.69.

LCMS (ESI+) *m/z*: 225[M + 1]$^+$, 248[M + Na]$^+$, 249[M + Na + H]$^+$.

Elemental Analysis: Calcd. For $C_8H_8BrN_3$; calculated: C, 42.50; H, 3.57; Br, 35.34; N, 18.59 Found: C, 42.45; H, 3.59; N, 18.62.

5-chloro-1-methyl-1H-indazol-3-amine (3d):

IR (KBr, υ_{max} in cm^{-1}): 3423 (N-H str., -Amine), 2976, 2943 (C-H asym. Str. aliphatic), 2834 (C-H sym. Str., aliphatic),3053 (C-H str., aromatic), 1541 (C=N str., Indazolyl), 1345 (C-N str.), 1501,1462 (C=C str., aromatic), 1114, 1122 (C-H i. p. def., aromatic), 863 (C-H o.o.p. def., aromatic),554 (C-Br str., Ar-Br).

^{1}H NMR δppm (DMSO-d_6): δ 3.81 (s, 3H, -CH_3), 5.36 (s, 2H, -NH_2),7.3–7.5(d, 2H, Ar-H), 7.85 (s, 1H, Ar-H).

^{13}CNMR δ ppm (DMSO-d_6): δ 35.41, 110.03, 114.34, 120.75, 125.88, 126.62, 136.59, 146.61.

LCMS (ESI+) *m/z*: 182[M + 1]$^+$, 205[M + Na]$^+$, 206[M + Na + H]$^+$.

Elemental Anal.Calcd. For $C_8H_8ClN_3$; calculated: C, 52.90; H, 4.44; N, 23.14 Found: C, 52.93; H, 4.40; N, 23.15.

4-methoxy-1H-indazol-3-amine (3e):

IR (KBr, υ_{max} in cm^{-1}): 3426 (N-H str., -Amine), 2983, 2943 (C-H asym. Str. aliphatic), 2834 (C-H sym. Str., aliphatic), 3053 (C-H str., aromatic), 1542 (C=N str., Indazolyl), 1346 (C-N str.), 1506,1460 (C=C str., aromatic), 1113, 1120 (C-H i. p. def., aromatic), 865 (C-H o.o.p. def., aromatic),553 (C-Br str., Ar-Br).

^{1}H NMR δ ppm (DMSO-d_6): δ 3.73 (s, 3H, $-CH_3$), 5.43 (s, 2H, $-NH_2$), 6.7 (d, 1H, Ar-H), 7.2 (d, 1H, Ar-H), 7.5 (d, 1H, Ar-H), 12.4 (s, 1H, -NH of ring).

^{13}C-NMR δ ppm (DMSO-d_6): δ 55.09, 95.02, 105.8, 106.8, 127.04, 142.02, 149.05, and 153.12.

LCMS (ESI+) *m/z*: 164[M + 1]$^+$, 187 [M + Na]$^+$, 188[M + Na + H]$^+$.

Elemental Anal. Calcd. For $C_8H_9N_3O$; calculated: C, 58.88; H, 5.56; N, 25.75 Found: C, 58.30; H, 5.43; N, 25.70.

6-bromo-1H-indazol-3-amine (3f):

IR (KBr, υ_{max} in cm^{-1}): 3421 (N-H str., -Amine), 2984, 2944 (C-H asym. Str. aliphatic), 2835 (C-H sym. Str., aliphatic), 3052 (C-H str., aromatic), 1541 (C=N str., Indazolyl), 1347 (C-N str.), 1504,1466 (C=C str., aromatic), 1117, 1125 (C-H i. p. def., aromatic), 866 (C-H o.o.p. def., aromatic),553 (C-Br str., Ar-Br).

^{1}H NMR δ ppm (DMSO-d_6): 5.38 (s, 2H, $-NH_2$), 7.4(d, 1H, Ar-H), 7.7(d, 1H, Ar-H), 7.8 (s, 1H, Ar-H), 12.4 (s, 1H, ring-N-H).

^{13}C-NMR δ ppm (DMSO-d_6): 107.04, 113.34, 120.05, 122.03, 124.22, 143.10, 149.32. LCMS (ESI+) *m/z*: 212[M + 1]$^+$, 235[M + Na]$^+$, 236[M + Na + H]$^+$.

Elemental Anal. Calcd. For $C_7H_6BrN_3$; calculated: C, 39.65; H, 2.85; N, 19.82 Found: C, 39.62; H, 2.82; N, 19.78.

5-iodo-1H-indazol-3-amine (3 g):

IR (KBr, υ_{max} in cm^{-1}): 3419 (N-H str., -Amine), 2984, 2945 (C-H asym. Str. aliphatic), 2836 (C-H sym. Str., aliphatic), 3055 (C-H str., aromatic), 1541 (C=N str., Indazolyl), 1343 (C-N str.), 1505,1467 (C=C str., aromatic), 1120, 1125 (C-H i. p. def., aromatic), 861 (C-H o.o.p. def., aromatic),556 (C-Br str., Ar-Br).

^{1}H NMR δ ppm (DMSO-d_6): 5.44 (s, 2H, $-NH_2$),), 7.4(d, 1H, Ar-H), 7.7(d, 1H, Ar-H), 8.2 (s, 1H, Ar-H), 12.4 (s, 1H, ring -N-H).

^{13}C-NMR δ ppm (DMSO-d_6): 88.3, 110.18, 115.37, 129.15, 134.23, 128.62, 140.31, 149.81 LCMS (ESI+) *m/z*: 259[M + 1]$^+$, 282[M + Na]$^+$, 283[M + Na + H]$^+$.

Elemental Anal. Calcd. For $C_7H_6IN_3$; calculated: C, 32.46; H, 2.33; N, 16.22 Found: C, 32.35; H, 2.36; N, 15.98.

5-fluoro-1H-indazol-3-amine (3 h):

IR (KBr, υ_{max} in cm^{-1}): 3425 (N-H str., -Amine), 2983, 2943 (C-H asym. Str. aliphatic), 2835 (C-H sym. Str., aliphatic), 3056 (C-H str., aromatic), 1546 (C=N str., Indazolyl), 1347 (C-N str.), 1502,1465 (C=C str., aromatic), 1119, 1126 (C-H i. p. def., aromatic), 865 (C-H o.o.p. def., aromatic),556 (C-Br str., Ar-Br).

^{1}H NMR δ ppm (DMSO-d_6): 4.98 (s, 2H, $-NH_2$),), 7.1(d, 1H, Ar-H), 7.6(d, 1H, Ar-H), 7.9 (s, 1H, Ar-H), 12.41 (s, 1H, ring -N-H).

^{13}C-NMR δ ppm (DMSO-d_6): 107.12, 110, 112.17, 115.17, 134.23, 149.81, 153.32.

LCMS (ESI+) *m/z*: 259[M + 1]$^+$, 282[M + Na]$^+$, 283 [M + Na + H]$^+$.

Elemental Anal. Calcd. For $C_7H_6FN_3$; calculated: C, 55.63; H, 4.00; N, 27.80 Found: C, 55.67; H, 3.98; N, 27.40.

3.2 Physico-chemical characterization

Benzonitrile (**1a-h**) is reacted with hydrazine (**2a-h**) in presence of catalyst ceric (IV) ammonium nitrate (CAN) in solvent EtOH-H_2O by using ultrasonic irradiation which undergoes cyclocondensation reaction to give substituted 3-aminoindazole (**3a-h**). The physical constants like melting point and solubility were determined for all the intermediate and final products. At every stage the reaction is monitored with TLC. Newly synthesized compound have been characterized on the basis of spectral data and elemental analysis such as FT-IR, ^{1}H NMR, ^{13}CNMR and mass spectra and they also screened for antimicrobial activities.

The IR spectrum of **3c** showed strong band absorption bands at 3424 cm^{-1} due to -NH- stretch in amine while bands at 1541 cm^{-1} is observed due to C=N stretch in Indazolyl and absorption band at 556 cm^{-1} showsC-Br aromatic stretching and stretch at 1344 cm^{-1} was observed for C-N str group confirms the cyclisation to form 5-bromo-1-methyl-1H-indazol-3-amine**3c**.^{1}H NMR of **3c** revealed a singlet signal at 3.65 ppm, at aliphatic region owing to three protons of -CH_3 group attached to aromatic ring, one more singlet at δ 5.47 ppm confirm protons of

	Zone of Inhibition (mm)											
Compd. Code	Gram + ve Gram –ve											
	S. aureus						*P.vulgaris*					
	Conc. (μg/mL)						Conc. (μg/mL)					
	1000	**500**	**250**	**125**	**63.5**	**31**	**1000**	**500**	**250**	**125**	**63.5**	**31**
3a	21	22	21	20	17	13	27	24	21	18	17	10
3b	23	21	20	18	16	18	25	22	19	16	14	09
3c	25	21	19	20	16	15	28	23	21	17	15	11
3d	23	22	21	19	18	15	27	24	20	18	17	12
3e	19	20	16	15	13	14	21	19	15	14	13	09
3f	24	23	21	20	17	16	25	23	21	19	15	08
3 g	23	20	18	17	16	13	23	22	21	16	14	10
3 h	20	18	16	15	13	11	26	21	17	15	12	09
DMSO	—	—	—	—	—	—	—	—	—	—	—	—
Std. Drug Chloramphenicol	25	22	20	19	17	15	26	24	23	21	17	15

Table 3.
Antibacterial profile of 3-aminoindazole derivative (3a-h) .

	Zone of Inhibition (mm)											
Compd. Code	Gram –ve											
	E. coli						*S.typhi*					
	Conc. (μg/mL)						Conc. (μg/mL)					
	1000	**500**	**250**	**125**	**63.5**	**31**	**1000**	**500**	**250**	**125**	**63.5**	**31**
3a	26	24	22	21	16	12	16	15	12	11	10	09
3b	24	21	19	18	17	16	16	14	10	09	06	07
3c	25	24	20	17	16	15	17	16	11	10	09	08
3d	26	23	21	20	17	11	14	10	08	08	07	09
3e	23	22	20	18	14	12	16	14	12	11	09	07
3f	25	24	22	20	18	15	17	16	11	10	09	08
3 g	21	20	17	16	14	11	15	13	11	10	08	07
3 h	22	21	20	18	16	13	14	12	10	08	06	05
DMSO	—	—	—	—	—	—	—	—	—	—	—	—
Std. Drug Chloramphenicol	26	24	23	21	17	14	17	15	12	11	09	08

Table 4.
Antibacterial profile of 3-aminoindazole derivative (3a-h).

–NH_2- group, Remaining ^1H NMR signal is present at aromatic region as expected. C^{13} NMR also supported data singlet signal at δ34.55, ppm revealed one aliphatic carbon, of -CH_3 attached to indazole ring. ESI-MS Mass spectra also confirm the molecular ion at *(m/z)* value at 225[M + 1]$^+$, and further supported by elemental analysis data found to be in good agreement with the molecular formula of $C_8H_8BrN_3$.

Potent antibacterial/inhibition profile of 3-amino indazole (at different concentration) by agar disc-diffusion method: - The entire novel synthesized heterocyclic compounds **3a-h** was screened for their *in-vitro* antimicrobial activity using disc-diffusion method. Their activity was compared with well-known commercial antibiotic Chloramphenicol. Antibacterial activity was determined by using Mueller Hinton Agar obtained from Hi media Ltd., Mumbai. Petri plates were prepared by pouring 10 mL of Mueller Hinton Agar for bacteria containing microbial culture was allowed to solidify. Test solutions were prepared with known weight of compound in DMSO and half diluted suitably to give the resultant

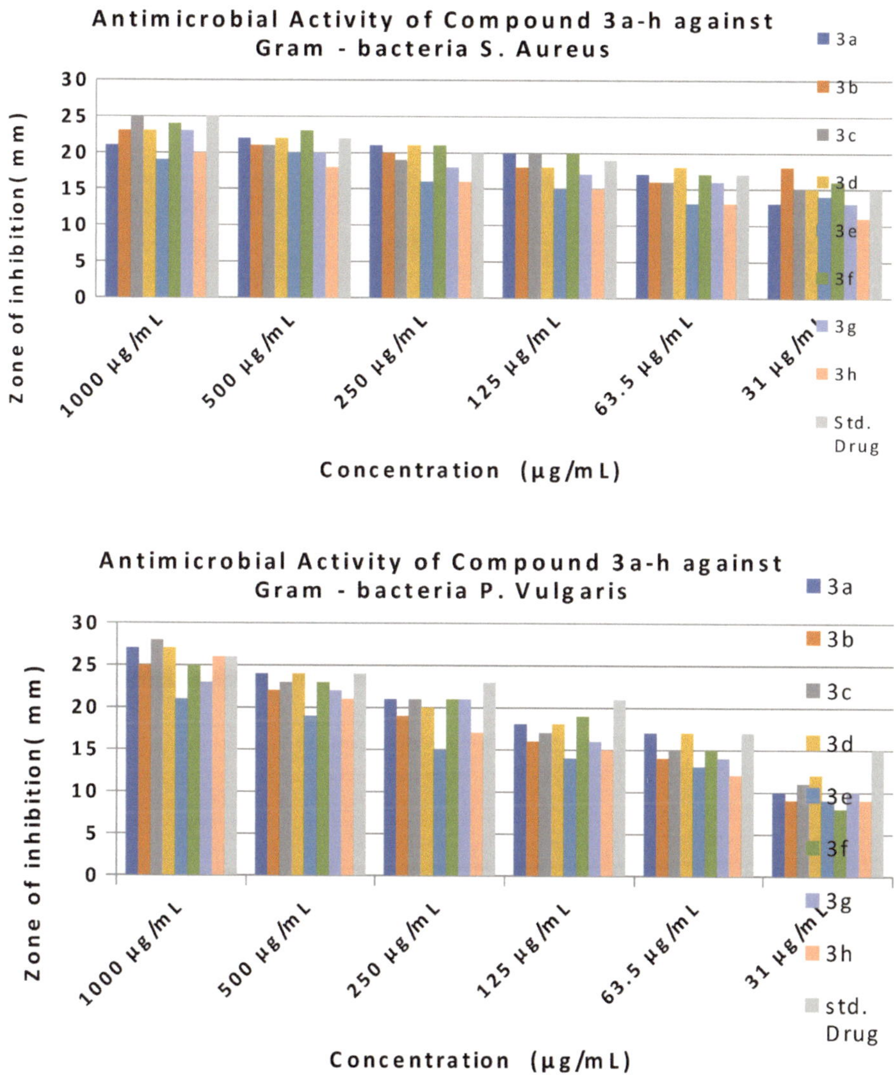

Figure 4.
Antimicrobial activity of compound 3a-h against Gram - bacteria S. aureus *and* P. vulgaris.

concentration of 31-1000 μg/mL. Whatmann no.1 sterile filter paper discs (6 mm) were impregnated with solution and allowed to dry at room temperature. The discs were then applied and the plates were incubated at 37°C for 24 h (bacteria) and the inhibition zone (**Figure 6**) was measured as diameter in four directions and expressed as mean. The results of the antimicrobial screening are illustrated in the **Tables 3** and **4**. From the results it is clear that the compounds tested showed variable toxicity against different bacteria.

3.3 Mechanism of inhibition/prohibition

Gram-negative bacteria habitually owe low susceptibility as outer membrane of their cell wall not gets blocked/penetrated by drugs easily and factors like amount of peptidoglycan, receptors, and lipids availability, nature of cross-linking, autolytic enzymes activity greatly influence the bio-activity, permeation, and incorporation of the antibacterial drugs. There are three behaviors in which antibacterial drugs have thus far beenshown to exert their definite actions upon bacteria: 1) by interfering with the synthesis of the relatively rigid cell wall which

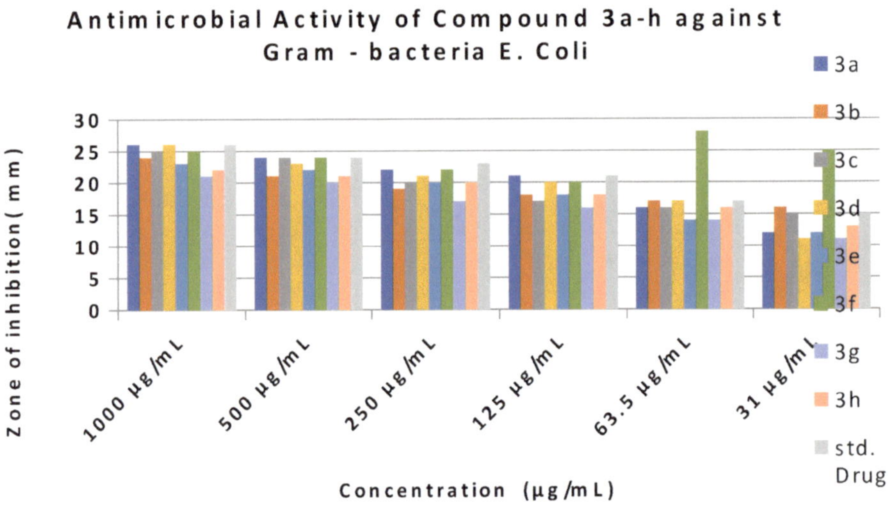

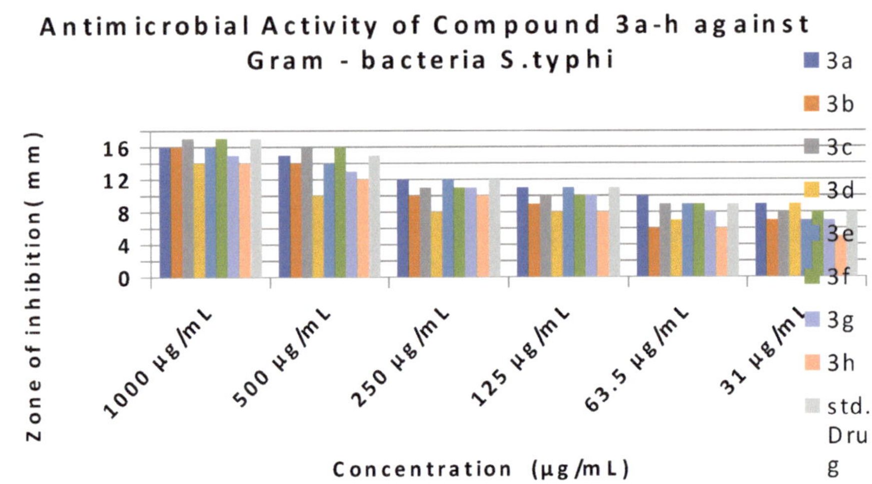

Figure 5.
Antimicrobial activity of compound 3a-h against Gram - bacteria E. coli *and* S. typhi.

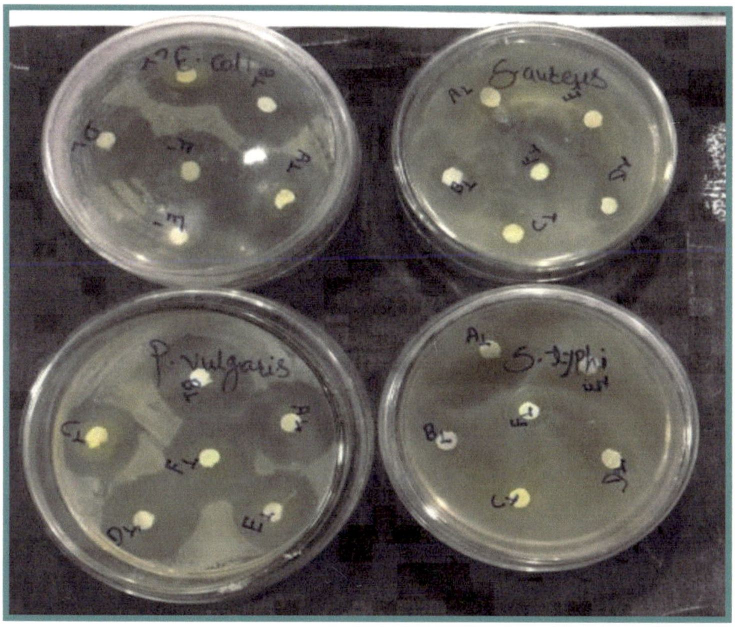

Figure 6.
Zone of inhibition in mm.

Maintains the structural integrity of the bacterial cell (2) by destructivethe elusive membrane, which encompasses the bacterial cytoplasm and assists as an essential osmotic barricade to the free diffusion of severalmetabolites and (3) by blocking cytoplasmic metabolic reactions which are involved in critical synthetic processes within the cell. Consequences of bioassay indicated that compound (**Figures 4** and 5) **3a, 3c, 3d, 3f** showed excellent activity against *S. aureus* and *P.vulgaris* and *E.coli*. Remaining compounds exhibited reasonable activity against selected bacterial strains.

4. Conclusions

In summary, in the present chapter we have described an efficient simple and reproducible method afforded various amino indazole derivatives (**3a-h**) in excellent yields, and without formation of undesirable side products. The synthetic protocol has been outlined in general reaction. At every stage the reaction was monitored with TLC. The physical constants like melting point and solubility were determined for all the intermediate and final products. Newly synthesized compound has been characterized on the basis of spectral data and elemental analysis such as FT-IR, ^{1}HNMR, ^{13}C NMR and mass spectra this chapter focused on helping futuristic researchers, clinicians, and academicians involved in synthesizing and corresponding biological screening of innate activity of certain novel amino indazole heterocycleswe have described here novel series of Indazole (**3a-h**) derivative. The presented series of compounds were synthesized in decent yields. The structure and purity of newly synthesized compounds were established by spectroscopic investigation and chemical examination. Antibacterial screening of synthesized substituted indazole (**3a-h**) derivatives exhibited a potent bactericidal. Thus, it could be powerfully stimulates major advances in remarkable significant chemotherapeutics in medicine, biology and pharmacy. Overall these indazole disturb macromolecules like cytoplasmic membrane covering cytoplasm which acts selective barrier to control internal composition of cell. Amino indazole derivative particularly interrupted such functional roles of cytoplasmic membrane and ionic outflow that resulted cell destruction/death. Synthesized potent bioactive substituted indazole derivatives may open new possibilities in the successful treatment of several diseases due to promising antibacterial profile. So, ample scope exists in further research of this heterocycle especially innate selectivity of these compounds needs to carry out their chemotherapy as potent antibacterial aims to target cell membrane of range of Gram-negative bacteria as to derive novel drugs ofInventive era. Among the synthesized compounds most of the compounds exhibited good to moderate activity against selected strains *S. aureus, P. vulgaris* and *E. coli* while poor activity was evaluated for *S. typhi* (**Figure 5**). This variation in toxicity it may attribute due to union ofdifferent substituent attached to the core 3-aminoindazole which may enhances the biological activities of parent nucleus.

Acknowledgements

The authors are gratified to, The Director, SAIF, Punjab University, and Chandigarh for providing, FTIR, CHN analysis, ^{1}HNMR, Mass Spectra, ^{13}C NMR. Authors are also thankful the Principal, Government Science College, Gadchiroli, for his support and cooperation and also pleased to Dr. Syed Abrar Ahmed and Dr. Mandar Paingankar, for there key assistance.

Conflict of interest

The authors declare no conflict of interest.

Author details

Chandrashekhar Devkate[1*], Satish Kola[2], Mohammad Idrees[3], Naqui J. Siddiqui[3] and Roshan D. Nasare[4]

1 Indraraj Arts, Commerce and Science College, Sillod Aurangabad, Maharashtra, India

2 M.G. Arts, Science and Late N.P Commerce College, Armori, Maharashtra, India

3 Department of Chemistry, Institute of Science, Nagpur, Maharashtra, India

4 Department of Chemistry Govt. Science College, Gadchiroli, Maharashtra, India

*Address all correspondence to: cgdevkate@gmail.com

References

[1] Yuxuan Y, Ilia K, Sheng F, Peng L, Stephen LB. Highly enantioselective synthesis of indazoles with a C3-quaternary chiral center using CuH catalysis. Journal of American Chemical society.2020;142: 10550.

[2] Tianshui Z, Weiliang B. Copper-catalyzed direct aryl quaternization of N-substituted imidazoles to form imidazolium salts Journal of Organic Chemistry. 2013;78:1317.

[3] Wansoo K, Hun YK, Kyungsoo O. Electrochemical radical–radical cross-coupling approach between sodium sulfinates and 2H-indazoles to 3-sulfonylated 2H-indazole. Organic Letters. 2020;22:6319

[4] Pan L, Chunrui W, Jingjing Z, Donald CR, Feng S. Synthesis of substituted 1H-indazoles from arynes and hydrazones. Journal of Organic Chemistry. 2012;77:3149

[5] Jie SZ, Clarabella JL, Ka YT, Niklas K, Jung-Ho S, Makhluf JH, Dean JT, Mark JK. Journal of American Chemical society.2019;141:6247

[6] Christian S, Steve K, John EM. New synthesis of 1-substituted-1H-indazoles via 1,3-dipolar cycloaddition of in situ generated nitrile imines and benzyne. Organic Letters. 2010;12:3368

[7] Chandrashekhar GD, Khandu DW, Digambar DG, Mohammad IMS. Eco-friendly synthesis of substituted indazoles from hydrazones. Journal of Chemistry and Chemical Sciences. 2015; 5:639

[8] Chandrashekhar GD, Khandu DW, Digambar DG, Mohammad IMS. Synthesis of substituted indazoles from N-tosylhydrazones using ionic liquid. Journal of Chemical and Pharmaceutical Research. 2015;7:533

[9] Jinyun D, Qijing Z, Zengtao W, Guang H, Shaoshun L. Recent advances in the development of indazole-based anticancer agents. ChemMedChem. 2018;13:1

[10] Arumugavel M, Koteswar RG, Bhabani SM, Duddu SS. Iron promoted C3–H nitration of 2*H*-indazole: direct access to 3-nitro-2*H*-indazoles. Organic and Biomolecular Chemistry. 2018;16: 5113

[11] Lars M, Wolfgang E, Nucleophilic carbenes and pseudo-cross-conjugated mesomericbetaines of indazole starting from analogues of the alkaloid-betainenigellicine. European Journal of Organic Chemistry.2005;10:2124

[12] M. Akram Khan. Chemical composition and medicinal properties of *Nigella sativa* Linn. Inflammopharmac ology. 1999;7:15

[13] Rob CW, Emma B, Ian BC, Simon JFM. A general one-step synthesis of substituted indazoles using a flow reactor. Organic Process Research and Development. 2011;15:565

[14] Han-Jun L, Shiang-Fu H, Chuan-Lin C, Mei-Huey L. A method for the regioselective synthesis of 1-alkyl-1H-indazoles.Tetrahedron. 2013;69:3907

[15] Xiaodong X, Yongwen J, Dawei M. Assembly of N,N-Disubstituted hydrazines and 1-aryl-1H-indazoles via copper-catalyzed coupling reactions. Organic Letters. 2012;14:2552

[16] Tuula K, Antti H, Noora K, Jan T, Robert F.A novel and efficient synthesis of 3-aminomethyl-N-tosyl-indazoles. Tetrahedron. 2010;66:8854

[17] Yajing L, Robert GB, Luke DL, Jonathan AE. Rhodium(III)-catalyzed indazole synthesis by C-H bond

functionalization and cyclative capture.2013;135:7122

[18] Manian RK, Ahbyeol P, Namjin P, Sunwoo L. Consecutive Condensation, C–N and N–N Bond Formations: A Copper- Catalyzed One-Pot Three-Component Synthesis of 2H-Indazole. 2012;13:3542

[19] Da-Gang Y, Mamta S, Frank G. Rh^{III}/Cu^{II}-Cocatalyzed Synthesis of 1*H*-Indazoles through C–H amidation and N–N bond formation. 2013;135:8802

[20] Fang S, Xian F, Xuan Z, Zhi-Bin H, Da-Qing S. An efficient synthesis of 2*H*-indazoles via reductive cyclization of 2-nitrobenzylamines induced by low-valent titanium reagent. Tetrahedron. 2012;68:3851

[21] Jiantao H, Yongfeng C, Yiqing Y, Yu R. A general and efficient approach to 2*H*-indazoles and 1*H*-pyrazoles through copper-catalyzed intramolecular N–N bond formation under mild conditions.Chemical Communications. 2011;47:10133

[22] Yao Z, Qiuling S. Oxidative ring-opening of 3-aminoindazoles for the synthesis of 2-aminobenzoates. Organic chemistry frontiers. 2018;5:3245

[23] Guihua C, Minglin H, Yungui P. Switchable Synthesis of 3-Substituted 1H-Indazoles and 3,3-Disubstituted 3H-Indazole-3-phosphonates Tuned by Phosphoryl Groups. The Journal of Organic Chemistry. 2018;83:1591

[24] Suleyman S, Z. Rustem A. And efficient synthesis of 3-aryl and 3-heteroaryl substituted a novel -1H-indazoles and their mannich derivatives. Synthetic communications. 2009;32: 3399

[25] Digambar DG, Archana DC, Chandrashekhar GD, Khandu DW, Amit PT, Rajendra PP, Abraham JD. Synthesis of indazole motifs and their medicinal importance: An Overview. European Journal of Medicinal Chemistry. 2015;90:707

[26] Raut SV, Tidke AD, Dhotre BK, Mohd AP. Synthesis of 3-[5-(substituted phenyl)-[1,3,4] oxadiazol-2-yl]-1H-indazole.Polycyclic Aromatic Compounds.2019;1:1

[27] Indazole derivatives and their therapeutic applications: a patent review (2013–2017). Ireen D, Sarel FM. Jacques J. Indazole derivatives and their therapeutic applications: a patent review (2013–2017). Expert opinion on therapeutic patents. 2018;28:441

[28] Galina NL, Emiliya VN, Valery NC, Oleg NC. Fluorine containing indazoles: Synthesis and biological activity. Journal of Flourine chemistry. 2016;192:1

[29] Chandrashekhar GD, Ajay MP, Khandu DW, Digambar DG, Mohammad IMS. Ceric (IV) ammonium nitrate catalyzed highly efficient synthesis of 3-aminoindazole and their antibacterial screening. International journal of research in pharmacy and chemistry.2017;7:513

[30] Neelam P Prajapati, Rajesh H Vekariya and Hitesh D Patel. Ceric Ammonium Nitrate (CAN)–Catalyzed Multicomponent Reactions: An Efficient Catalyst for Green Organic Synthesis. Synthetic Communications. 2015;45: 2399–2425.

[31] Ahmed K, Shaikh F, Shaheer MM, Moku B, Shaik B, Rafique M, Siddiqui H, Abdullah A. Convenient synthesis of substituted pyrroles via a cerium (IV) ammonium nitrate (CAN)-catalyzed Paal–Knorr reaction. Arabian Journal of Chemistry. 2016;9:542

[32] Vellaisamy S, J. Carlos M. Cerium (IV) ammonium nitrate as a catalyst in organic synthesis. Chemical Reviews. 2010; 110:3805

[33] Cailin AB, Eunbyeol K, Spencer C, Mahalingam B, Bryan RG, Nirala S. Structures and free energies of cerium ions in acidic electrolytes. Inorganic Chemistry. 2020;59:12552

[34] Juan FG, Damiano R, Tomas T, Pedro M, J. Carlos M. One pot synthesis of functionalized carbazoles via a can-catalyzed multicomponent process comprising a C–H activation step. The journal of organic chemistry. 2017;82: 7492

Section 4

Green Computing Industry and Applications

Chapter 6

Green Computing: A Machinery for Sustainable Development in the Post-Covid Era

Wilson Nwankwo and Paschal Uchenna Chinedu

Abstract

Socio-economic sustainability has emerged the common song of the policy makers globally. It has been projected as a developmental strategy by international and regional agencies. There has been several campaigns and programs all of which are intended to promote sustainability. In developing countries, the masses have been bamboozled with often unrealistic bogus policies hypocritically crafted in a bid to deceive the uninformed who are undoubtedly helpless in the midst of the conundrum. However, the 2019 reports of the IPCC and OECD respectively on global warming, sustainability and climate change is not a phenomenon that should be swept under the carpet by any sensible government. Though in many jurisdictions, campaigns and policies have long assumed political undertone, it must be stressed that it is time for talking the walk. Governments must put up implementable strategies that are all encompassing across the various sociopolitical classes and the different industry levels. According to the said reports, global warming and climate change pose severe challenges to sustainability and this is attributed to social, and economic root causes. The social sources are conflicts and poor socio-political governance structures whereas the economic sources are connected to *industry, electricity, residential, agriculture, and transport. It is reported that 60% of greenhouse emissions globally emanate from the economic source.* The worst hit is the sub-Saharan Africa where the dumping of electronic wastes and uncontrolled deployment of unregulated hardware for industry operations have remained a major environmental menace in the last decade. Having regard to the foregoing, this paper seeks to provide a systematic inquiry into the green computing policies and legislations in a major economic hub in the sub-Saharan Africa. The essence of this investigation is to critically review the present status of existing policies, strategies, and legislation vis-à-vis their strengths, lapses, and the contributory effect of these on driving the sustainability programs in the general developmental outlook of the sub region.

Keywords: green computing, environmental computing, computer wastes, climate change, sustainability

1. Introduction

Green computing (GC) has remained an important aspect of vital discussions involving environmental protection, green energy, climate change and sustainable development, though several equivalent terms such as Green ICT (GICT), ICT sustainability, and perhaps in a broader terminology, 'environmental computing', one thing remains glaring that is, the common goal which, according to the global body on GICT, the International Federation of Global and Green Information Communication Technology (IFGICT), is to practice and achieve eco-friendly deployment and use of ICT in the society. Ordinarily, the green computing ideology is aimed at reducing the perilous component load during computing equipment manufacturing. It is also concerned with energy-efficient and eco-friendly computing as well as processes for marshaling out machineries that enhance the biodegradability of computer-based wastes [1]. Green computing cover the spectrum of large-scale computing environment such as data centres, and mobile systems [2, 3]. It has been noted that green computing is a major leap in mitigating environmental pollution, degradation, and the impacts of climate change [1]. Adopting the tenets of green computing does entail producing and using energy-efficient computing devices, developing cutting edge strategies and research towards reducing energy consumption, and the appropriate disposal of electronic waste arising therefrom. Abugabah and Abubaker proposed a 5-phase lifecycle strategy for green computing [4]. According to the said proposal, green computing encompasses green design, green production, green procurement, green operations, and green disposal. Whereas the proposition is apt, it must be noted that these technical phases are not naturally continuous as in a process cycle but could occur independently. Consequently, each phase may present unique concerns that are influenced and shaped by social and economic forces (e.g. policies, legislations, lifestyles, culture, etc.) hence must be factored in for an effective greener ICT ecosystem. With the ever-growing global population, the demands for computing and ICT devices may experience sustained increase and the consequence is enormous pressure on the manufacturers to either release new technologies or increase the volume of production of current brands [5]. There is no gainsaying that technology advancement has its downsides or side effects which include generation of e-wastes, pollution, health and environmental degradation. In this chapter, three key factors that promote these side effects have been identified. They are: Unregulated and indiscriminate deployment of heavy ICT infrastructure; rapid technology evolution; and socioeconomic inequities in developing countries.

1.1 Unregulated deployment of energy-intensive ICT infrastructure

The unregulated deployment of heavy ICT infrastructure such as data centres and base stations for telecommunications, is a major contributor to environmental pollution and other environmental imbalances. Nwankwo and Ukhurebor noted that data centres have become indispensable in driving socioeconomic activities globally in recent times [6]. Data centres are not only run by technology service providers (TSPs) such as telecommunications and ICT service provisioning companies but are commonplace in the industry sector (cement factories, steel production factories, consumer goods production industries, automobile factories, finance and banking organizations, etc.) of every economy. These energy-intensive infrastructure are localized in many industrial zones in developing and developed countries. It has been noted that across all economic sectors including the public sector, modernization programs are not complete without ICT infrastructure [7–12]. Research has shown that data centres contribute significantly to greenhouse emissions, pollutions and eventually climate change [13–20]. In the post-COVID era, there is likely to be a surge in the deployment of data centres, mobile stations, and other sophisticated computing equipment as the pandemic is engendering an era wherein at least 80% of all economic activities are to run online. This has been christened the 'new normal' across different jurisdictions. With this predicted surge, Governments must put up implementable strategies and machineries that are all encompassing across the various sociopolitical classes and the different industry levels.

1.2 Rapid technology evolution and unfair competition

Rapid technology evolution and unfair competition globally, is a serious contributory factor to the menace of electronic dumping especially in Africa and other developing jurisdictions [21]. Over the years, Africa has emerged the targeted destination for sales of computing hardware from Europe, Asia, and America regardless of the product specifications in relation to standard, safety, and ecofriendliness [21, 22]. Atkin described this anomaly as an environmental injustice occasioned on the developing nations by the developed countries (Europe, United States, Japan, Korea, Australia, China, etc.) who intentionally transport tons of e-waste to vulnerable countries though with adequate knowledge that these poor countries do not have the resources to dispose these wastes [23].

Interestingly, despite Africa's economic prowess and abundant human and natural resources that would have promoted manufacturing or local assembly of computing equipment within the continent, it appears the continent is sabotaged by those countries with supposedly superior technical and technological endowments. This inequity suffered on a large scale by the continent across the global economic sphere, is one of the factors that occasioned the continued inflow of substandard and electronic wastes into the continent amid regional and national campaigns against electronic dumping.

As noted by [24] electronic dumping has remained one of the most worrisome environmental issue throughout the African continent. According to their study, second-hand and discarded electronic products from Europe, Asia and America is very predominant and has continued to create an uneasy atmosphere for the disposal machineries and strategies put in place by the various national governments. According to them the computer components that are popularly shipped to Africa are categorized into three: those that have reached end of life but bought by some Africans and other nationals that trade in second-grade goods; products phased out from mainstream distribution and support, by their manufacturers; and used products already discarded by their owners in foreign countries.

Several researchers have linked electronic dumping to advancement in technology and the increasing output in the production of computing and electronic devices in western countries and Asia [25, 26]. As more industries across different jurisdictions churn out new devices, the existing ones though had not reached their end-of-life may be discarded regardless of their utility values owing to the release and promotion of newly launched sophisticated models. A typical example is mobile phones and personal computers (PCs). These two devices are emerging part of everyday life in the socioeconomic ecosystem as technology rapidly becomes the vehicle for social and economic activities. The existing traditional and/or simple devices are being re-engineered using Artificial Intelligence (AI) and Internet of Things. Notice that these two pervasive technologies are rapidly evolving into full-fledged consumer-oriented technologies. The quest for intelligence in devices is exerting a tremendous effect on the development of new devices that would in no time replace the existing ones. In other words, the major concern would no longer be the end of life of the equipment per se but the utility value. Thus, where the utility of the device is found relatively lower vis-à-vis that of the emerging device, then it would be discarded in favor of the more sophisticated device thereby adding to the global e-waste burden. Currently, it has been observed that semiconductor devices and sensors are being added to products that were never before contemplated to have such components in them. For instance, the desire for wearable monitors, smart city, smart homes, smart agriculture, intelligent TV, etc. with embedded capabilities to exchange information with other devices, though laudable, also contributes immensely to the problem of electronic wastes.

It has also been observed that the newly produced devices often exhibit shorter life spans owing to the use of non-removable batteries. Typical examples are smart phones, tablets, consumer health monitoring devices, etc. A decade ago, almost every smart phone or tablet has a battery which could be easily replaced at will by the owner of the device once the battery's performance falls below a certain range. The trend is different nowadays. Currently, these devices come with non-removable batteries. The implication is that once the batteries malfunction or get exhausted, the devices themselves would be considered useless and would ultimately be discarded and new ones acquired.

1.3 Socioeconomic inequities in developing countries

The quest for survival is the major promoter of the growing informal recycling business in developing countries. It is reported that the supposed boom in informal and unregulated recycling of failed and discarded computing equipment is associated with the ongoing massive shipping and dumping of these systems in developing countries. According to [22] these computer wastes are intentionally shipped to developing countries especially in Africa with the aim of selling them to users who may put them to use or recycling to recover some valuable elements (gold, copper, etc.) which may be sold thereafter to meet economic demands [27]. It is reported that these businesses thrive in countries such as China, Nigeria, Ghana, India, the Philippines, Thailand, Vietnam, etc. [28–31]. The downside of such businesses is the adoption of traditional methods of recycling that often release toxic and hazardous substances into the environment [32, 33]. Again, the massive production of these devices is connected to the rate at which they become obsolete. It is reported that in 2016 alone, about 49 million tons of electronic wastes were generated and the said report predicts an increase to 57 million metric tons of e-wastes in 2021 [34]. Another report states that the global e-waste burden stood at 5.8 kg per person in 2014 which rose to 6.3 kg per individual globally in 2017 [35]. According to the global e-waste monitor, 53.6 million tons of e-waste was generated globally in 2019 [36]. This report agrees perfectly with the projection made by Cho in 2018 [34]. On average, 40 million tons of electronic waste are generated globally every year [37]. According to a report, the United States alone disposes over 47.5 million computer systems and hundred (100) million cell phones among other electronic wastes each year [23]. It is estimated that proper disposal of a ton of e-waste would cost 2500 USD in a developed country [23]. Developing countries rarely have these resources for e-waste disposal. Despite this, these countries allow imports of e-wastes at 3 USD per tonne.

These findings call for urgent measures especially as human existence is increasingly confronted with more health challenges such as the Covid-19 pandemic.

The trend in e-waste generation has emerged a serious health and sustainability issue globally [37, 38]. For over two decades, there has been a continuous engagements of several global and regional organizations including the various agencies of the United Nations, African Union, European Union, national agencies for environmental protection, and non-governmental agencies. These efforts have been directed towards developing strategic interventions that would enable humanity deal with the peculiar and lasting challenges occasioned by e-waste. Some of the notable engagements include:

a. The Libreville Declaration on Health and Environment in Africa: this was held in 2008. It was the premier Inter-Ministerial Conference (IMC) to consolidate the pledges and declarations on environment, health, and safety [38];

b. The Busan Pledge for Action on Children's Environmental Health [37, 39]. This was an offshoot from the third conference on Children's health and the Environment by the World Health Organization (WHO) held in 2009 at Busan, Korea

c. The UN Decade on Ecosystem Restoration which adopts a 9-point strategy to restore the degrading global natural ecosystem [40]

d. United Nations Framework convention on Climate Change [41]

Amid the calls from different quarters in respect of climate change and environmental protection especially as it affects e-waste management it appears the ongoing programs and initiatives have adopted a collective approach in the sense that electronic wastes is a term that is all encompassing. Accordingly, some distinctions are important as not all electronic wastes emanate from computing devices and not only computer devices contribute to pollution and environmental degradation. It therefore follows that a particularized approach which would x-ray all the vital areas of computing deployments and applications taking into consideration the entire computing device forms and lifecycle (production, acquisition, deployment, use, withdrawal, and destruction/recycling) and their contributions to sustainability domains such as environmental protection, safety, and health. The destruction/recycling phase is a critical point in the lifecycle of electronic products generally. It is also the most demanding phase. Direct destruction through burning and deposition into landfills create sustainability problems. The by-products of burning are pollutants and usually toxic to humans and in some cases plants. Though recycling e-waste is promising, however, with the present poor recycling facilities across Africa, weak policies and regulations, and lack of recycling programs [38], the continent is in dire need of green ICT reforms. The aim of this chapter is to showcase the relevance of green computing on fostering sustainability through the design and entrenchment of mechanisms and approaches including policies and legislations that would forestall the impending danger that might be occasioned on humanity by the continuous accumulation of computer-based wastes, and the deployment of energy-intensive computing facilities.

2. Implications of computer-based wastes and energy-intensive computing

2.1 Computer-based wastes

Ordinarily, computer-based wastes are often in the solid form as opposed to liquid and gaseous wastes from other sources (see **Figure 1**). Examples of such wastes include: Computers including PCs, high end server hardware, Telephones and smart phones, Network devices (routers, switches, gateways, radios, etc.), Chips, Base stations, Motherboards, Printers, Wireless devices, Fax and copiers, Cathode ray tubes and Monitors, and Transformers. However, due to the physicochemical complexity of the components that constitute these electronic products, the tendency to generate liquid and gas emissions are very high. Typically, materials used to produce computing hardware contain several reactive elements such as lead, silicon, silver, mercury, platinum, copper, cobalt, palladium, aluminum, cadmium, lithium, selenium, etc. According to a report by the environmental protection agency of the United States [23], while comparing the amount of certain elements contained in the e-wastes and that mined from raw ores indicated that it is estimated that the amount of gold in a ton of electronic circuit boards is 40–800 times more than that from the ore, and that the quantity of copper in one ton is 30–40 times more than the quantity of mined copper from a metric ton of raw ore. With these reactive elements in electronic dumps, the underlying health implications of these components are brought to bear.

Figure 1.
Computer-based e-wastes for sale in Ikeja Lagos Nigeria.

It is interesting to note that computer-based wastes like other e-wastes contain toxic heavy metals such as mercury, lead, cadmium, beryllium, plastic (polyvinyl chloride), and hazardous chemicals (e.g. brominated flame retardants) that are harmful to the health of individuals in particular and the environment generally. It has been observed that most of these wastes are shipped to developing countries for use and possible recycling [21, 42, 43]. These wastes produce gaseous emissions that pose health risks to the individuals within and around such e-wastes are dumped or recycled [42]. Having recognized the dangers associated with such wastes, countries such Nigeria has made necessary legislative provisions to guard against such hazardous or harmful wastes. Section 15 of the Harmful Wastes (Special Criminal Provisions etc.) Act, rightly provides that "harmful wastes depicts any injurious poisonous, toxic or noxious substance and, in particular, nuclear wastes emitting any radioactive substance ... as to subject a person to the risk of death, fatal injury or incurable impairment of physical or mental health". The effect of this provision is that any waste whether or not electronic that could cause some harm is a harmful waste. There is no gainsaying that wastes from computing equipment could be hazardous enough to pollute the air, water, and soil. The contamination of these three environmental layers is akin to humankind intentionally creating a hazardous ecosystem antithetical to sustainable development and growth. These hazardous e-wastes could route toxic chemicals through the soil, air, water thereby providing the channel for generalized environmental degradation and pollution that promotes health anomalies and imbalances such as infections, respiratory stress, allergic reactions, visual impairment, poisoning, hematological problems, etc. (See **Table 1**). It has been reported that these wastes have strong nexus with adverse birth outcomes, thyroid dysfunction, behavioral changes, lung failure, and adverse cellular changes [30], Kidney failure, cancer, etc. **Table 1** shows some toxic metals and compounds from computer wastes and their health implications.

2.2 Energy-intensive computing (EC)

EC involves the deployment and utilization of energy-intensive computing equipment especially to drive socioeconomic activities. Data centres globally fall within the ranks of energy-intensive computing. They are not only common in developed countries but are also at the cornerstone of industry in developing countries. The intensive use of data centres and high capacity computing hardware are popular from the telecommunications subsector to banking and finance, manufacturing and production, agricultural and food processing, mining and extraction, and educational institutions. EC infrastructure popular in mission critical applications such as in manufacturing, production, mining, telecommunications, and the financial services subsectors. EC is a creator of both computer wastes and pollution. Despite

Metal/ Compound	State	Where used in computing hardware	Health implications	Route
Cadmium [44–46]	Solid. Its pyrolysis generates toxic fumes	Resistors, Nickel-cadmium batteries, Screens, lasers	Kidney, lung, bone damage, muscle pain, chills, fever, nausea, vomiting, decreased memory and cognition abilities, impaired neuro-motor skills	Inhalation, Ingestion
Lead [30, 37]	Solid. Lead oxide is formed when burnt/heated	CRT monitors, printed boards, polyvinyl chloride formulations	The mucous membranes within the lungs, skin, abdomen, easily absorb lead into the blood which may predispose the individual to lead poisoning, high blood pressure, liver and kidney damage, reduced nervous development and permanent nervous damage in children	Inhalation, ingestion
Mercury	Liquid/ Gaseous	Fluorescent tubes and flat screen monitors.	Whether alone or in methylated forms, this substance is harmful to the digestive, nervous and immune systems respectively. The lungs, kidneys and other internal organs are affected. Mercury salts are corrosive to the eyes, skin and the gastrointestinal tract	Inhalation Ingestion
Beryllium oxide	Solid/Liquid	Thermal grease for heat-sinks; Processor, power transistors, vacuum tubes.	Causes irritation skin, throat, nose, and lungs.	Inhalation Contact
Barium	Solid/Liquid	CRT monitors	Poisoning	Inhalation Ingestion, contact
Zinc [30]	Solid	CRTs, Metal coatings	Poisoning, vomiting, nausea, diarrhea, cramps	Inhalation, ingestion

Table 1.
Some health implications of computer/ICT wastes.

the economic importance of EC infrastructure in driving industrialization, they are noted as potential causes of greenhouse emissions (GHG), pollution, and agents of climate change [15–19]. The gases contribute to warmer climate can affect the ecosystems locally and globally [47]. For instance, extreme weather affects agricultural crop production and yield, livestock production, desertification, health challenges including epidemics, ocean acidification, heavy precipitation, flooding, food supply and security challenges, new diseases, and energy supply problems [27, 48, 49].

3. Challenges confronting green computing

Green computing is confronted with many challenges. These are categorized into: Manufacturer-induced bottlenecks, Economic challenges, Policy, Consumer attitude, and Social exploitation.

3.1 Manufacturer-induced bottlenecks

The manufacturer-induced bottlenecks are a major barrier to the adoption, implementation and enforcement of green computing practices and policies. The equipment manufacturer is a major player in the distribution and supply business chain. In the local parlance, the equipment manufacturer is tainted as having 'the fork and the knife' in the sense that the equipment manufacturer is technically the addition to influencing the life span of computing equipment, the manufacturers are in principle aware of how long a device would be supported or maintained. Generally, these manufacturers play a major role in determining when a device becomes obsolete. These they may do by intentionally modifying the design, embedded software, or even withdrawing their support for specific models of the device. They often project the benefits of the newly produced variant of the device in terms of cost and functionalities while at the same time hyping the costs of maintenance of the older models. Experience has shown that many of such claims do not reflect the reality rather the manufacturer's desire to maximize profits at the expense of the consumer and the environment. Another area of concern is the use of materials that are not ecofriendly in the synthesis and production of electronic goods. If green manufacturing is strictly adhered to, green computing challenges would be significantly reduced [50–53]. In the same vein, the power consumption requirements by devices may depend on the manufacturer's specifications and readiness to use ecofriendly and sustainable materials [52]. It therefore follows that manufacturers can drastically reduce power demands by carefully selecting and using materials that require less energy to operate. However, [52] note that manufacturers face challenges of incurring high cost and low returns in their bid to switch to greener technologies, green production and manufacturing processes, compared to the traditional methods. Their study also revealed that these companies encounter difficulties in generating the required energy by greener and/or renewable methods.

3.2 Economic challenges

Inequities in the distribution of resources is a major factor in the battle against electronic dumping and green technologies. As has been adumbrated by researchers, developing countries are grossly disadvantaged in the management of e-wastes. This may be connected to the poor economies, political will and leadership failure [54]. The developed countries often exploit these gaps while dealing with the developing nations. Due to indebtedness to the developed nations, many developing nations lack the willpower to reject such exploitation. As survival is the ultimate desire of every individual, small and medium scale businesses involving in buying and selling of second-grade electronic products is common. It is also not uncommon to see large businesses who are involved in the importation of these second-hand computing equipment and by extension e-waste. For instance, Alaba market located in the Ojo area of Lagos Nigeria is popularly known in the sub-Saharan Africa for its exploits in deals involving second-hand computing equipment from Europe and America. The same is true for the popular computer village in Lagos and Agbogbloshie in Accra Ghana [55]. In Lagos, these small and medium Majority of these thousands of businesses involving the importation of and trade in fairly used computing products reflect the inequities occasioned on the masses by the leadership and economic maladies. It is important to stress that despite the ban on the importation of some of those goods in developing countries such as Nigeria, such goods are either smuggled or concealed in other cargoes during shipping and inspection of those cargoes at the destination port is often fraught with a lot of irregularities including corruption currently owing to human intervention which has remained a major problem in the inspection and clearance processes at the ports [56–58]. With the rising spate of unemployment and no remedy in sight, it is believed that trade in contraband goods would always thrive despite the policies and regulations in place. There, the inflow of e-waste is very much likely going to abate in continental Africa till the various governments implement functional policies and machineries to reduce economic hardship and unemployment.

3.3 Policy challenges

Like developing countries, majority of developing countries have policies, legislation, guidelines, etc. associated with environmental protection [6]. Often, the problem in developing countries may not be the legislation or policy per se but the machinery to realize the provisions contained by the policy. Where these machineries are not properly harnessed gaps would continue to widen [6]. Several pitfalls connected to the operations of these machineries have been noted ranging from poor and crude infrastructure, incoherent initiatives of government operatives, poor maintenance culture, corruption, nepotism, social stratification with the emergence of the 'untouchables' that include individuals and organizations which are covertly above the law, poor enforcement, to negligence and abdication of responsibilities on the part of the personnel manning the public agencies. Two major problems confronting majority of policies including legislations and guidelines for driving GC ideals are noncompliance and weak enforcement [6, 59, 60]. Noncompliance and weak enforcement had linked to sociopolitical and economic problems [47, 61, 62].

3.4 Consumer attitude

Consumer attitude is a major factor in the supply and demand chain. In developing countries, the demand trend especially for electronic products such as smartphones and other mobile computing devices does not often reflect the true status of the economies. Consumers regardless of their social statuses often desire to have sophisticated devices. It is common to see a student in a secondary school with two or more smart phones. In recent times ownership of such devices has become a denominator for social class differentiation and imposition among people. Consequently the drive to upgrade to latest technologies is a common behavioral disposition among the young and the old. There is need for a public re-orientation if this problem is to be nipped in the bud otherwise it would continue to promote energy overload and environmental menace as more devices are discarded on a daily basis.

3.5 Social exploitation

Social exploitation involves taking advantage of the lapses in the policies, rules, guidelines, and legislations as well as customer's ignorance and demand, by manufacturers and importers of computing devices. For instance, in societies where there are no clear laws applicable to enforcement of green computing ideals, importers and manufacturers are bound to exploit this gaps in shipping all manner of products without recourse to anything.

4. Solutions to green computing challenges

Going forward, the following points are stated as steps towards resolving the challenges confronting green computing globally.

a. International and National Policies on Green Computing including trans-border movement restrictions and enforcement which would necessitate mutual understanding between countries within a region. Adoption of regional taskforces to police the movement of computer-based products regardless of their condition. Review of trans-border customs guidelines and entrenchment of stricter inspection at border posts, airports, and sea ports. Initiating a global ban on the export and import of second-hand computing devices would be an ultimate global policy to control the menace of electronic waste movements generally.

b. Strict regulation on manufacturers: Computing device manufacturers are the biggest player in the fight against dumping. Policies on the use of eco-friendly and biodegradable materials in the manufacturing process cannot be overemphasized.

c. Creation of vibrant machinery for consumer awareness programs, consumer protection and enforcement of consumer rights including the right to repair, disassemble and replace parts of malfunctioned equipment. This would relieve the manufacturers the pressure of providing support for its equipment over a long period.

d. Establishment of global grants for the development of electronic recycling industry subsector. E-waste disposal is somewhat a difficult enterprise. Recycling is plausible means of generating economically viable products from the wastes, reducing environmental pollution and promoting healthy environment. However, in the last decade, informal recycling is predominant in developing countries.

e. Continued advocacy for the adoption of green production and manufacturing practices i.e. the use of environmentally friendly and sustainable materials

f. Design and Standards advocacy: Green computing should be developed as a standalone international standard and maintained by the international organization for standardization (ISO) just like other plethora of standards it defines and sustains. Once internationalized, governments can domesticate and adopt them into their mainstream economic policies, guidelines, and regulations.

g. Promotion of circular economy [63, 64]. A circular economy has the potentials of mitigating against the challenges posed by dumping and poor manufacturing practices. Entrenching a circular economy implies that there would be no waste in the lifecycle of the computing equipment. In other words, appropriate machineries and policies are put in place to ensure compliance across board. Every product produced by a manufacturer would be recyclable once it reaches its end of life or ungracefully discarded. The producers and other recycling facilities are readily accessible to consumers. Thus collection, dismantling, refurbishing, re-use, and recycling of e-waste are formally implemented in the producer-consumer experience cycle.

h. Adoption of technology-driven architecture that would employ a resource planning system to support collective and collaborative management of risks associated with computer wastes and the adoption of multi-platform waste management systems including bioremediation, biomining, etc. [65, 66]. The proposed architecture would extend any existing practices through the deployment of intelligent computerized approaches that includes risk and hazard profiling of various industry subsectors, industry operators, location of operation, environmental impact etc.

5. SD and GC in the post-covid era

SD is connected to GC and greener ICT. The post-covid period is an era of uncertainties especially in public health and safety ecosystem [67, 68]. This calls for a review of and the strengthening of the existing measures, policies, legislations, guidelines and controls on SD. As noted by [68], in the post-covid era, environmental degradation and climate change have higher tendency of causing more terrifying ecological catastrophes. According to [69] social inclusion, environmental sustainability and economic growth are the major objectives of SD. GC is a major driver of SD because its ideals, practices, and principles focus on a cleaner, safer, and sustainable environment that promote the total wellbeing of all members in the natural ecosystem. **Table 2** summarizes the relationship between green computing and SD. The 17 SD goals are presented I sequence with the projected contribution of green computing practices stated against each goal.

S/N	SD goal	Contributions of GC practices
1	No poverty	GC would ensure cleaner environment in the workplace which would guarantee the productivity, health, and safety of the average worker.
2	Zero hunger	GC would reduce soil pollution and contamination thereby promoting better crop yields and food security
3	Good health and Well-being	Green houses gases from computing devices and e-waste threaten the health and wellbeing of humans. If such threats are eliminated, there would be improvement in the social wellbeing of the people.
4	Quality education	Learning, research, and teaching are all affected positively by a clean and safe environment. Education under such circumstances produces knowledgeable experts who would contribute to national and global development
5	Gender equality	GC would eliminate water, air and soil contamination as well as good health. With cleaner environment and water women and girls would be relieved of the roles they play in homes such as searching for water and cleaning of the environment
6	Clean water and sanitation	Proper disposal of e-wastes and control of carbon load from large computing environments would prevent leaches that pollute underground water
7	Affordable and clean energy	GC advocates renewable and clean energy for computing equipment and allied devices
8	Decent work and economic growth	Safer and cleaner work environment motivate workers and this would boost productivity and by extension economic growth.
9	Industry, innovation and infrastructure	GC is an innovative ideology that promotes pollution-free industrial environment and harmony. It advocates for the development of sustainable devices and infrastructure
10	Reducing inequality	Pollution governance and actions can ensure that no group or community bears a disproportionate share of the harmful effects of pollution.
11	Sustainable cities and communities	Safe, clean and non-toxic environment promote sustainable production and economic activities in local and urban areas
12	Responsible consumption and production	GC drives sustainable consumption and manufacturing activities by encouraging the adoption of greener manufacturing and production processes by manufacturers which would reduce the incessant demand of newer devices by consumers owing to the fall in the utility value of existing devices
13	Climate action	GC promotes use of renewable energy devoid of high carbon loads; and the clean disposal of computer-based wastes which could pollute the environment thereby adding to the climate change burden
14	Life below water	Toxic wastes from electronic dumps contaminate underground waters leading to the death of essential aquatic organisms. GC advocates for proper e-waste disposal

S/N	SD goal	Contributions of GC practices
15	Life on land	One of the main goals of GC is safe, healthy and clean environment at all times
16	Peace, justice and strong institutions	Enforcement of GC policies and regulations without prejudice would promote peace and safety across all socioeconomic spheres thereby attracting confidence and trust for public institutions
17	Partnerships for the goals	GC ideal and policies should operate at national, regional and international levels. Such collaborations would engender unity of purpose and maximum impact as to the control of electronic dumping as well as the enforcement of greener production practices. The result would affect all spheres of the ecosystem

Table 2.
How GC practices promotes SD goals.

6. Conclusions

This chapter examines the green computing domain and articulates the implications of advances in the development and deployment of computing and ICT infrastructure globally laying emphasis on the growing spate of electronic dumping in the sub-Saharan Africa where countries like Nigeria and Ghana are greatly affected. It identifies some of the health and environmental problems which GC is intended to eliminate or reduce. Though it makes case for the adoption of the green computing ideals to ensure the realization of SDGs [69] in the post-covid era, however, it contemplates various barriers and challenges to the adoption of green computing ideals and proffers various feasible solutions that could help eliminate these challenges. In conclusion, it is established that GC is a viable machinery for sustainable development through a one-to-one mapping of SD goals to the goals of green computing.

Conflict of interest

There is no conflict of interest in this work.

Author details

Wilson Nwankwo[1*] and Paschal Uchenna Chinedu[2]

1 Informatics and Cyber Physical Unit, Department of Computer Science, Edo University, Iyamho, Edo State, Nigeria

2 Department of Computer Engineering, Edo University Iyamho, Edo State, Nigeria

*Address all correspondence to: nwankwo.wilson@edouniversity.edu.ng

References

[1] Nwankwo W, Olayinka AS, Ukhurebor KE. Green Computing Policies and Regulations: A Necessity? International Journal of Scientific and Technology Research.2020: 9(1).

[2] Curry E,Guyon B, Sheridan C, Donnellan B. Developing a Sustainable IT Capacity: Lessons from Intel's Journey. MIS Quarterly Executive. 2012; 11(2):61-74.

[3] Mittal S. A survey of Techniques for Improving Energy Efficiency in Embedded Systems. IJCAET. 2014; 6(4): 440-459.

[4] Abugabah A, and Abubaker A. Green computing: Awareness and practices. 2018 4th International Conference on Computer and Technology Applications (ICCTA), Istanbul, 2018: 6-10. doi: 10.1109/CATA.2018.8398646.

[5] Sharma M, Joshi S, Kumar A. Assessing enablers of e-waste management in circular economy using DEMATEL method: An Indian perspective. Environ SciPollut Res Int. 2020; 27(12):13325-13338. doi: 10.1007/s11356-020-07765-w. Epub 2020 Feb 4. PMID: 32020449.

[6] Nwankwo W and Ukhurebor KE. Data Centres: A Prescriptive Model for Green and Eco-Friendly Environment in the Cement Industry in Nigeria. International Journal of Scientific and Technology Research. 2020; 9(5):239-244

[7] Basalisco B, Hansen MBW, Haanperä T,Dahlberg E,Hansen MM, Brown J, Münier LL, Laursen MF, Næss-Schmidt HS. Google's Hyperscale Data Centres and Infrastructure Ecosystem in Europe: Economic impact study. Copenhage Economics, 2019

[8] Pinto S. New data centers in Africa aim to drive economic growth [internet]. Available at https://channels.theinnovationenterprise.com/articles/new-data- centers-in-africa-aim-to-drive-economic-growth. Accessed [2020-11-20]

[9] ESI Consult Solutions. The Economic and Revenue Impact of Data Centers in Pennsylvania: An analysis of the potential growth of data centers in PA as a result of the proposed data center sales and use tax exemption [Internet]. ESI. 2019. Available from https://econsultsolutions.com/wp- content/uploads/2019/06/Data- Center-Final-Report_Econsult-Solutions.pdf. Accessed [2020-11-10]

[10] Gonçalves PVR. The economic impact of data centres[Internet] 2018. Available from https://www.diplomacy.edu/blog/economic-impact-data-centres. Accessed [2020-11-20]

[11] Government of Ireland. Government Statement on The Role of Data Centres in Ireland's Enterprise Strategy. Department of Housing, Planning and Local Government. 2018

[12] Thornton G. A Study of the Economic Benefits of Data Centre Investment in Ireland. IDA Ireland. 2018

[13] Anthony B, Majid MA, Romli A. A Descriptive Study towards Green Computing Practice Application for Data Centers in IT Based Industries. MATEC Web of Conferences. 2018: 150:1-8. doi: 10.1051/matecconf/201815005048

[14] Singh M, Sidhu AS.GREEN COMPUTING. International Journal of Advanced Research in Computer Science. 2016; 7(6) (Special Issue): 195-197, 2016.

[15] Kurp P. Green computing. Communications of the ACM. 2008; 51(10):11-18

[16] Baylis S. Green computing. ECMWF Newsletter. 2011; 126:28-31

[17] Andreopoulou ZS. Green Informatics: ICT for Green and Sustainability. Agricultural Informatics. 2012; 3(2):1-8

[18] Martinez-Fernandez C, Hinojosa C, Miranda G. Greening Jobs and Skills: Labour Market Implications of Addressing Climate Change. OECD Local Economic and Employment Development (LEED) Papers 2010/2. OECD Publishing.2010.

[19] Kadam MS, Singh D. Approaches to Green Computing to Reduce Global Warming. IBMRD's Journal of Management and Research. 2013;2(1):413-420

[20] Pinto SM, Divya V, Varsha R ,Nalina V. Green Computing and Energy Consumption Issues in the Modern Age. International Journal of Engineering and Techniques. 2018; 4(3): 661-665

[21] Sajid M, Syed JH, Iqbal M, Abbas Z, Hussain I, Baig MA. Assessing the generation, recycling and disposal practices of electronic/electrical-waste (E- Waste) from major cities in Pakistan. Waste Manag. 2019; 84:394-401.

[22] Weiss TC. Electronic Waste: Medical and Health Issues [Intenet]. Disabled World. https://www.disabled-world.com/health/ewaste.php

[23] Atkin M. The Environmental Injustice of Electronic Waste [Internet]. 2019. Available from http://chej.org/2015/06/19/the-environmental-injustice-of- electronic-waste/. Accessed [2020-11-20]

[24] Olayinka AS, Nwankwo W, Olayinka TC, and Osiele MO. Design of Automated Power Outlet Distribution System: Recycling Electronic Wastes. International Journal of Advanced Research in Computer and Communication Engineering. 2018: 7(9).

[25] Hossain MS, Al-Hamadani SMZF, Rahman MT. E-waste: A Challenge for Sustainable Development. Journal of Health and Pollution.2015:5 (9): 3-11.

[26] Bhutta MKS, Omar A, and Yang X. Electronic Waste: A Growing Concern in Today's Environment. Economics Research International.2011:2011:1-8

[27] Wang K, Qian J, Liu L. Understanding Environmental Pollutions of Informal E- Waste Clustering in Global South via Multi-Scalar Regulatory Frameworks: A Case Study of Guiyu Town, China. Int J Environ Res Public Health. 2020; 17(8):2802.

[28] Brune M-N, Goldizen FC, Neira M, van den Berg M, Lewis N, King M, Suk WA, Carpenter DO, Arnold RG, Sly PD. 2013. Health effects of exposure to e- waste. Lancet Global Health 1:e70.

[29] Ghisellini P, Cialani C, and Ulgiati, SA. Review on circular economy: the expected transition to a balanced interplay of environmental and economic systems. Journal of Clean. Production. 2016: 114, 11-32.

[30] Grant K, Goldizen FC, Sly PD, Brune M-N, Neira M, van den Berg M, Norman RE. Health consequences of exposure to e-waste: a systematic review. Lancet Global Health. 2013; 1(6): e350-e361.

[31] Guerrero, LA, Maas G, and Hogland W. Solid Waste Management Challenges for Cities in Developing Countries. Waste management, 2013: 33(1), 220-232.

[32] Li W, and Achal V. Environmental and health impacts due to e-waste disposal in China – A review. Science of The Total Environment.

2020; 737. https://doi.org/10.1016/j.scitotenv.2020.139745.

[33] Gollakota ARK, Gautam S, Shu CM. Inconsistencies of e-waste management in developing nations - Facts and plausible solutions. J Environ Manage. 2020; 261:110234.

[34] Cho R. What Can We Do About the Growing E-waste Problem? 2018

[35] Research and Markets. Global E-waste Recycling & Reuse Services Market Size, Market Share, Application Analysis, Regional Outlook, Growth Trends, Key Players, Competitive Strategies and Forecasts, 2018 To 2026. 2019. Research and Markets.

[36] Forti V, Baldé CP, Kuehr R, and Bel G. The Global E-waste Monitor 2020: Quantities, flows and the circular economy potential. United Nations University (UNU)/United Nations Institute for Training and Research (UNITAR) – co-hosted SCYCLE Programme, International Telecommunication Union (ITU) & International Solid Waste Association (ISWA), Bonn/Geneva/Rotterdam.

[37] WHO. Inheriting a sustainable world? Atlas on children's health and the environment. 2017. Geneva

[38] WHO. Libreville Declaration on Health and Environment in Africa [Internet]. 2008. Libreville. https://www.ehrn.co.za/download/libreville_declaration.pdf. [Accessed: 2020-10-10]

[39] WHO. 3rd International Conference on Children's Health and the Environment. 2009

[40] United Nations. The United Nations Decade on Ecosystem Restoration [Internet]. 2020. Available from https://wedocs.unep.org/bitstream/handle/20.500.11822/31813/ERDStrat.pdf?sequen ce=1&isAllowed=y. Accessed [2020-11-10]

[41] United Nations. Framework Convention on Climate Change. 2015. Geneva

[42] Heacock M, Kelly CB, Asante KA, Birnbaum LS, Bergman ÅL, Bruné MN, Buka I, Carpenter DO, Chen A, Huo X, Kamel M, Landrigan PJ, Magalini F, Diaz-Barriga F, Neira M, Omar M, Pascale A, Ruchirawat M, Sly L, Sly PD, Van den Berg M, Suk WA. E-Waste and Harm to Vulnerable Populations: A Growing Global Problem. Environmental Health Perspectives. 2016 May; 124(5):550-5.

[43] Heacock M, Trottier B, Adhikary S, Asante KA, Basu N, Brune MN, Caravanos J, Carpenter D, Cazabon D, Chakraborty P, Chen A, Barriga FD, Ericson B, Fobil J, Haryanto B, Huo X, Joshi TK, Landrigan P, Lopez A, Magalini F, Navasumrit P, Pascale A, Sambandam S, AsliaKamil US, Sly L, Sly P, Suk A, Suraweera I, Tamin R, Vicario E, Suk W. Prevention-intervention strategies to reduce exposure to e-waste. Rev Environ Health. 2018 Jun 27; 33(2):219-228.
[38] Ramzan S, Liu C, Munir H, Xu Y. Assessing young consumers' awareness and participation in sustainable e- waste management practices: a survey study in Northwest China. Environ SciPollut Res Int. 2019 ; 26(19):20003-20013.

[44] Bernhoft RA. Cadmium Toxicity and Treatment. The Scientific World Journal. 2013; 2013:1-7. https://doi.org/10.1155/2013/394652

[45] Attila H, Erdei S, Horvath G. Comparative studies of H2O2 detoxifying enzymes in green and greening barley seedlings under cadmium stress. Plant Science. 2001; 160: 1085-1093

[46] Sharma A, Sachdeva S. Cadmium Toxicity And Its Phytoremediation A Review. International Journal of Scientific & Engineering Research. 2015;6(9):395-405

[47] Nwankwo W, Ukhurebor KE. An X-ray of Connectivity between Climate Change and Particulate Pollutions. Journal of Advanced Research in Dynamical and Control Systems. 2019; 11(8) Special Issue: 3002-3011.

[48] Bakshi B, Nawrotzki RJ, Donato JR, Lelis LS. Exploring the Link between Climate Variability and Mortality in Sub-Saharan Africa. Int. J. Environment and Sustainable Development. 2019; 18(2): 206-237.

[49] Ukhurebor KE, Aigbe UO, Olayinka AS, Nwankwo W, Emegha JO. Climatic Change and Pesticides Usage: A Brief Review of Their Implicative Relationship. AU-Ejournal of Interdisciplinary Research. 2020; 5(1): 44-49.

[50] Baines T, Brown S, Benedettini O, Ball P. Examining green production and its role within the competitive strategy of manufacturers. Journal of Industrial Engineering and Management. 2012; 5(1):53-87

[51] Rehman MAA, Shrivastava RL. Green manufacturing (GM): past, present and future (a state of art review). World Review of Science, Technology and Sustainable Development (WRSTSD). 2013; 10(1/2/3)

[52] Karurkar S, Unnikrishnan S, Panda SS. Study of Environmental Sustainability and Green Manufacturing Practices in the Indian Automobile Industry (June 30, 2018). OIDA International Journal of Sustainable Development. 2018; 11(6):49-62

[53] Eshikumo SM, Odock SO. Green Manufacturing and Operational Performance of a Firm: Case of Cement Manufacturing in Kenya.International Journal of Business and Social Science. 2017; 8(4)

[54] Ngum E. Letters: The price of poor political leadership in Africa [Internet]. The Standard. 2019. Available from https://standard.gm/letters-the-price-of- poor-political-leadership-in-africa/#:~:text=Poor%20political%20 leadership%20in%20Africa%20 has%20contri buted%20to,crisis%2 0with%20corruption%20at%20 all%20levels%20of%20government. Accessed[2020-12-01]

[55] Orisakwe OE, Frazzoli C, Ilo CE, Oritsemuelebi B. Public Health Burden of E- waste in Africa. J Health Pollut. 2019 Jun 4;9(22):190610.

[56] Nwankwo W, Olayinka AS, Benson BU. X-ray Cargo Scanning and Risk Management in Trade Facilitation: Analysis and Model of an Online Imaging and Documentation Management System, International Journal of Modern Education and Computer Science.2019:11(5).

[57] Nwankwo W. Customs automation: The X-ray and Computerized Risk Management Systems era. International Journal of Science and Research. 2017; 6(4).

[58] Nwankwo W, Olayinka AS. Implementing a risk management and X-Ray cargo scanning document management prototype, International Journal of Scientific and Technology Research. 2019;8(9).

[59] Nabegu MAB, Naibbi AI. Environmental Regulations in Nigeria: A Mini Review. International Journal of Environmental Sciences & Natural Resources. 2017; 1(5):1-3

[60] Akamabe UB, Kpae G. A Critique on Nigeria National Policy on Environment: Reasons for Policy Review. IIARD International Journal of Geography and Environmental Management. 2017; 3(3):22-36.

[61] Gana AJ, Ngoro D. An Investigation into Waste Management Practices in Nigeria (A Case Study of Lagos

Environmental protection Board and Abuja Environmental protection Board). West African Journal of Industrial & Academic Research. 2014; 12(1): 122-126.

[62] Ijaiye H, Joseph OT. Rethinking Environmental Law Enforcement in Nigeria. Beijing Law Review 2014; 5:306-321.

[63] Ferronato N, Rada EC, Gorritty Portillo MA, Cioca LI, Ragazzi M, Torretta V. Introduction of the circular economy within developing regions: A comparative analysis of advantages and opportunities for waste valorization. J Environ Manage. 2019 Jan 15; 230:366-378.

[64] Singh A, Panchal R, Naik M. Circular economy potential of e-waste collectors, dismantlers, and recyclers of Maharashtra: a case study. Environ SciPollut Res Int. 2020 Jun; 27(17):22081-22099. doi: 10.1007/s11356- 020-08320-3.

[65] Ramos M, Machón-González I, Garcia H, Calvo-Rolle J. Visual supervision of a waste water biological reactor using artificial intelligence algorithms. 2013; 1-6. 10.1109/SmartMILE.2013.6708192.

[66] Sinha S, Kaur G. Potential applications of advanced biosensor systems for the real-time monitoring of wastewater treatment plants In S.K. Brar, K. Hegde, and V.L. Pachapur (Eds), Tools, Techniques and Protocols for Monitoring Environmental Contaminants. 2019; 75-94. Amsterdam, Elsevier.

[67] Lancaster K, Rhodes T, Rosengarten M. Making evidence and policy in public health emergencies: lessons from COVID-19 for adaptive evidence-making and intervention. Evidence & Policy: A Journal of Research, Debate and Practice. 2020

[68] Mustafa B. The age of uncertainty and the post-COVID-19 era [Internet]. Economy and Politics. Available from https://wsimag.com/economy-and-politics/62139-the-age-of-uncertainty-and-the-post-covid-19-era. Accessed [2020-12-01].

[69] Nwankwo W. and Njoku CC. Sustainable Development in Developing Societies: The Place of ICT-driven Computer Education. International Journal of Emerging Technologies in Learning. 2020; 15(12).

Chapter 7

The Environmental Influence of Tax Regimes in Selected European Union Economies

Fortune Ganda and Rufaro Garidzirai

Abstract

Eurostat and the European Environmental Agency have in 2019 reported there is still need to continue implementing zero-carbon practices in European Union (EU) Countries although there has been a noted decrease of 22% in emissions when compared to their 1990 levels. This paper employed a system-Generalised Method of Moments (GMM) framework to evaluate the environmental impacts of tax systems in selected 28 EU economies from 2010 to 2017. The results of the study proved that aggregate environmental tax is not effectively lowering greenhouse gas emissions as expected, although it improves environmental sustainability. Possibly the environment tax revenue collected in the European Union countries was not used to enhance energy efficiency; hence it could not lower greenhouse gas emissions. The other findings demonstrate that when environmental tax is disaggregated (energy tax and transport tax) these instruments have been more efficient in lessening emissions and also improves environmental sustainability (in the case of transport tax). The paper, therefore, highlights the importance of adopting green tax instruments which are more focused and harmonising directly with environmental goals for EU economies.

Keywords: greenhouse gas emissions, environmental sustainability, energy tax, transport tax, environmental tax, eco-innovation rating, production scores, green research and development, government expenditure, economic growth

1. Introduction

Conversations on emission and its effects on the economy and environment are increasing especially in developed countries. Among the key issues surrounding discussions on emission is the impact of taxation on carbon emission and environmental sustainability. There is unanimity among researchers on the effect of emissions on the environment. That is to say, emissions cause environmental degradation, diseases, reduces household welfare and are detrimental to economic growth and development [1–3]. In this light, it is evident that climate change has become a global problem [4]. This global problem has awakened the need for governments worldwide to invent techniques to minimise environmental issues and emissions. Some of the methods include subsidies, ecological laws, taxes, environmental policies and awareness programs [5]. Of particular importance are taxes which are an integral instrument in dealing with emission [6]. The significance of taxation on emission and environmental degradation has captured the attention of researchers and policymakers in developing policies and recommendations on minimising emission.

There is a noticeable increase in taxation on emissions and environmental sustainability in the European Union (EU). Taxation takes the form of energy tax, environment and transports tax especially, in Slovenia, Poland, France, Portugal, Finland, Latvia, Ireland and Denmark [7]. The main purpose of these taxes is to minimise emissions up to an acceptable level of 5 percent [8]. Furthermore, the government introduced environmental and emission taxes to reduce negative externalities caused by third parties in production and consumption since nobody takes responsibility for creating them [9]. The negative externalities include pollution, land degradation and the greenhouse effect that tends to cause diseases, low standards of living, low quality of products, reduction in income and energy consumption [10]. Since firms do not take any responsibility, the EU has taken the responsibility of reducing emissions to an acceptable standard [11]. Thus, fiscal authorities have imposed a certain amount of $50–100 per ton on production for any environmental misuse and emission [11]. Tax experts argue that the $50–100 per ton tax that is not shown in the final price of goods and services covers the social costs suffered by the third party [2]. Furthermore, environmental and emission taxes increase government revenue and contribute to economic growth significantly. A survey conducted by Sterner and Kohlin [12] found that environmental and emission tax contributes 8 percent of the government revenue and 3 percent of economic growth in the European Union region.

The introduction of environment and emission taxes has sparked heated debates among scholars. The main crust of these debates is whether taxation is an effective way of reducing environmental emissions. Noteworthy is the complementary school of thought which contends that tax on environmental emission addresses market failures to an acceptable standard and reduces health diseases [3, 8]. This means that taxation on emissions brings about efficiency and effectiveness in the production of goods and services since firms get to develop new regulations that foster efficiency and reduce the cost of production. Also, environmental tax improves the quality of the products produced in production processes [13]. On the other, the substitutive school of thought argues that taxes on emission are inclined to fiscal policies rather than environmental policies [14]. The substitutive school of thought recognises that environmental and emission taxes focus more on raising government revenue than reducing emissions as such taxes tend to be regressive as prices of goods and services change [15]. In this sense, the substitutive school of thought concludes that taxation does more harm than good because it causes a greater degree of the loss of welfare as compared to emission. Therefore, this study envisaged contributing to the current debate on tax on emissions and environmental sustainability.

Central to the problem is that environmental pressures are a global phenomenon. The European Union is not exempted from this problem. Environmental pressures have become an issue of concern as the emission threat has increased over the past years. Another factor that has become a cause for concern is the forecast by economists that emissions are likely to increase to 35 percent, and this poses a threat to environmental sustainability [16]. Of importance is that these environmental pressures pose a risk to people's health, welfare, and economy. Astuti and Maryono [17] note that emissions cause health diseases such as eye irritation, asthma, and pneumonia. Emissions and environmental pressures do not only undermine the environment and health faculties but affect the economic operations of a country as well. There is no doubt that these challenges should be addressed. The economic theory prescribes many methods of solving these challenges. Such methods include environmental tax, fuel tax, awareness programs, subsidies to mention but a few. Relevant to this study is environment taxation and tax emission, which are the main focus of this study. Hence, this study investigated the influence of emission on selected EU countries.

The purpose of the study is to contribute to the existing literature on environmental accounting significantly. Most of the studies have focused on the effects of tax on carbon emission [2, 6, 18, 19]. However; this study takes a different stance by examining the environmental influence of tax structures in selected EU economies by taking into account both short-run and long-run dynamics. The study focused on other variables that are immensely important to environmental issues and yet are barely used by other researchers. These variables include research and development, production scores, eco-innovation ratings and the different types of tax such as energy, transport, and environmental tax. The authors of this study conducted a thorough search of the relevant literature. They found no study that combined all the variables in one study to investigate the environmental influence of tax structures in selected EU economies. Hence, the current paper covers this research gap to find robust results that are important to policymakers. Furthermore, the results and the nature of the research provide a niche for future researchers focusing on few limitations of the study. The study also contributes to the body of existing knowledge on natural environmental studies.

Therefore, this paper is organised as follows. The literature review is summarised in the next section. The methodology and variables used in the study are discussed in Section 3. This also includes the source of the data and prior expectations. The empirical results and analysis are discussed in Section 4, while Section 5 consists of the summary, conclusion, recommendations, and limitations of the study.

1.1 Environmental tax in the European Union

The introduction of environment tax in the European Union can be traced back to 1990 [20]. Since then, it has received attention from various governments intending to minimise environmental degradation. The idea was to charge polluters a certain fee per unit of the damage they have caused to third parties. In line with this objective, the European Union introduced four types of environment tax, namely energy, transport, pollution and resources tax [12]. For the purposes of this study, researchers focused on energy tax and transport tax as they are widely used in the European Union. Eurostat [21] defined energy tax as a certain amount paid by the energy sector for causing negative externalities. Energy tax target polluters who make use of petrol, diesel, biofuels, electricity consumption and carbon fuels [21]. The energy tax is mainly used in Italy, Germany, Netherlands, France, Sweden and Finland as they use heavy power plants and consumes much electricity compared to other countries [22]. This also implies that these countries receive more tax revenue from electricity tax while Sweden and Denmark get more revenue from fuel tax. Second is the transport tax, which is an amount paid for making use of vehicles and vehicle ownership [23]. It includes the importation of motor vehicles, flight tickets, toll gates, car registrations and insurances [21]. This form of tax was introduced to raise revenue and minimise greenhouse gas emissions. The European Commission [24] reports that 25 percent of greenhouse gas emissions are caused by the transport sector of which road transport contributes 75 percent to these transport emissions followed by civil aviation and navigation respectively. This type of emission is common in Norway, Netherlands, Finland, Greece, Spain and Denmark.

Noteworthy is that energy tax is widely used in the region to reduce greenhouse gas emission. There has been an increase in the use of tax that leads to an increase in the energy tax revenue since 2000–2018. The increase in revenue has also led to the rise in the Gross Domestic Product. The contribution of environmental tax to GDP was experienced a decade later after the introduction of the environment tax in 1990. Notably is the 5 percent contribution from 2013 to 2019. Despite the use of environmental tax, the Eurostat [21] found contrasting results in the European Union. On one hand, it is a significant increase in greenhouse emission in countries such as Germany, France and Italy. On the other hand, it is a significant decrease in greenhouse emission in countries such as Lithuania, Latvia and Romania [25]. From the discussions, the environment tax influences greenhouse gas emission differently in the European Union individual countries depending on the environment policies used by each country towards eradicating the environment hazards. The question still remains: Does environmental tax reduces greenhouse emission and improves environmental sustainability since its implementation is influenced by price elasticity of energy and transport demand?

2. Literature review

This section is divided into two parts. The first part examines how tax and other variables influence carbon emissions in selected economies. The second part evaluates how tax frameworks affect environmental sustainability in countries studied.

2.1 Empirical literature on carbon emission

The influence of emission tax cannot be separated from past studies on taxation, economy and environmental economics. For instance, a survey carried out by Miller and Vella [19] investigated whether taxes are effective in dealing with pollution. The study examined if taxes on emission help to produce quality products. The study targeted 50 countries across all the regions and used panelised dynamic regression models. The results of the study revealed that taxes reduce carbon emission in all the countries. Also, the study showed that the quality of products is improved if the polluters are taxed. Similarly, Metcalf [6] achieved the same results that carbon tax reduces carbon emissions in Britain, Columbia and the United States of America. Metcalf [6] further observed that taxation on emission improves employment and economic growth. Worthy of note is that the preceding studies present a negative relationship between carbon tax and carbon emission. Thus, taxation on emission is the most effective way of reducing emissions to an acceptable level. The studies concur that taxes reduce environmental pollution despite their difference in geographical location.

In South Africa, carbon tax also has an inverse relationship with emission. This result was concluded by [2] who examined the effects of carbon tax on the economy. The study employed the dynamic Computable General Equilibrium modeling methodology and found an inverse relationship between carbon tax and emission. The study further showed that carbon tax is negatively related to economic growth. Thus, the more firms pay carbon tax the fewer goods and services they produce, thereby compromising economic growth. Klier and Linn [26] concur with these results as they reach the same conclusion after using the panel regression analysis in Sweden, France, and Germany. The authors' objective was to investigate the relationship between vehicle carbon taxation and carbon vehicle emission. This relationship was prevalent in France compared to other countries. Since firms were taxed for emission, a decrease in emissions from vehicles was experienced in all the countries. The common denominator between these two studies is that taxation has a negative effect on economic growth despite the use of different methodologies and geographical locations. A salient point to note on the carbon tax is that it discourages firms to be innovative and this leads to a decrease in investment and eventually a decrease in economic growth.

Lin and Li [18] using a panel regression analysis, examined the impact of carbon tax on carbon emission in selected European countries. The authors found three sets of results: a negative relationship between carbon tax and emission in Finland; a positive relationship between carbon tax and emission in Norway and no relationship was identified in Netherlands, Denmark, and Sweden. Since Norway is one of the heavy carbon polluters in Europe, taxing the firms reduced emission. The same result was achieved by Di Cosmo and Hyland [27] who concluded that carbon tax is an effective way of reducing emissions in Norway. On the other hand, in the Netherlands, Denmark and Sweden carbon tax did not influence carbon emission. This result is contrary to the findings of Lin and Li [18] who found an inverse relationship between carbon tax and emissions. The authors further propounded that fiscal authorities should increase tax on emitters for carbon tax to be effective. Moreover, an interesting result is the positive relationship between carbon tax and carbon emission found in Norway. This result is not common in the Organization for Economic Co-operation and Development (OECD) since all the governments joined hands to reduce emissions through the Piogiouvot method.

Anderson [28] inquired whether a carbon tax is the solution to greenhouse emissions. The author used 11 European Union countries as his case study. To achieve the aim of the study, the author employed a quasi-experiment and found that tax curtailed emissions by 11 percent. The study confirmed the economic theory that prescribes that carbon tax deals with negative externalities whilst also reducing emissions. A similar result was found in Norway by Bruvoll and Larsen [29]. The authors employed simulations and a diverse index from 1990 to 1999 and the study revealed an emission reduction of 2.3 percent. Revoredo-Giha et al. [30] examined the impact of carbon taxes on greenhouse emissions in the United Kingdom. The study showed that carbon tax reduces greenhouse emissions. Gonzalez [31] and Haites [32] concur with the above-mentioned studies by reiterating that carbon tax is the best instrument to reduce greenhouse emissions and the most effective approach in reducing emissions.

Concerning the relationship between economic growth and emission, Ameyaw and Yao [33] analysed the impact of economic growth on carbon emission in West African countries from 2007 to 2014 using panel regression. The results show an unidirectional cause from GDP to carbon emission. Thus, an economy that taxes emissions is likely to improve economic growth. The same result was also achieved by Asongu et al. [34] who investigated carbon emissions and economic growth and found a relationship running from economic growth to carbon emission to energy consumption. An interesting result was found by [35] who examined the effects of economic growth on emission in developing countries. The study used panel analysis and found a negative relationship between economic growth and emissions while [36] found no link. The study examined the link between energy consumption, emissions and economic growth in the Middle East and North Africa (MENA). The rationale behind this finding is that taxation discourages firms to produce more goods and services due to increased cost of production.

Other authors emphasised the fact that carbon tax on emission is regressive in nature and leads to loss of welfare [14, 15]. For instance, Devarajan et al. [15] found that taxes on carbon emission reduce household welfare by 40 percent, whilst also reducing carbon emission by 15 percent. In other words, carbon tax works better in reducing household welfare than in minimising emission, its main objective. The study further found that carbon tax is regressive as poor households spend more than 50 percent of their salaries on taxed goods and services. This result was also found by [14]. Marx and Slamang [37] and Sterner [38] examined the relationship between energy and carbon tax on emissions in European Union countries. The study concluded that transport taxes, energy and carbon taxes are regressive.

2.2 Empirical literature on environmental sustainability

Liobikiene et al. [39] investigated the role of energy taxes on climate change in the European Union. The main focus was to check if environmental tax influences environmental sustainability. The authors applied panel data methods and found that environmental tax influences environmental sustainability in a positive way. The same results were found by Nerudova et al. [40] who examined the tax system and environmental sustainability in the European Union and found a positive relationship between the two. Park and Yoon [41] studied the link between environment tax and sustainable development in China, Japan and Korea using a survey. The study revealed a positive relationship between the two in all these countries. It seems the above-mentioned studies point to a positive relationship between taxation and environmental sustainability. Thus, taxation on environmental pollution improves environmental sustainability. A study by Radulescu et al. [42] in Romania employed the Ordinary Least Square and Vector Error Correction Model. The authors found a negative relationship between environmental sustainability and environmental tax. The authors argued that fiscal authorities should use other methods other than taxation to achieve environmental sustainability.

Streimikiene et al. [43] added economic growth as a variable that was not examined by [39, 42] by investigating the role of green tax on sustainable energy development in Baltic countries. The study found a positive relationship between environmental tax, economic growth and environmental sustainability. The authors propounded that taxation ensures environmental sustainability that has a direct influence on economic growth. Kurniawan and Managi [44] and Moisesca [45] arrived at the same conclusion by examining the relationship between economic growth and sustainable development in Indonesia from 1990 to 2014. The study used the inclusive wealth framework and found that economic growth influences environmental sustainability in a positive way. From all the studies that examined the link between economic growth and environmental sustainability, a positive relationship was achieved therein.

Urbaniec [46] conducted a study on eco-innovation and environmental sustainability. The main objective was to assess the role played by eco-innovation on environmental sustainability. The study concluded that eco-innovation minimises environmental damage. Similar results were also concluded by [47] who carried out a study on the role of eco-innovation and environmental sustainability in Malaysia. The findings of both studies point to the fact that an increase in environmental compliance improves the environment. Another common denominator is that both studies used the same methodology: the theoretical structural model and found similar results. The eco-innovation is also positively linked to Research and Development, thus Powe [48] found that research and development have a positive impact on environmental sustainability. The authors argued that Research and Development yields results in big sectors, while in small sectors a link was not found. The same results were also found by [49] who examined the green economy and sustainable development worldwide from 2002 to 2010. The study found that research and development have a positive impact on environmental sustainability. However, Sauvé et al. [50] found a negative relationship between Research and Development and environmental sustainability. The authors arrived at this conclusion after employing an ordinary least square first difference.

Kim and Yoon [51] examined the relationship between environmental sustainability and production in manufacturing firms. The objective of the study was to check the impact of production on environmental sustainability. The results reveal that production has a positive influence on environmental sustainability. The same results were also found by [52] who examined environmental sustainability and production. The preceding results differ from those achieved in the study done by [53] who examined the relationship between sustainable environment and production using the trend and content analysis. The study found that production causes environmental hazards. The author concluded that production in developing countries over utilises resources with the objective of combating poverty. On the other hand, production in developed countries over utilises resources for export purposes.

Saud et al. [54] examined the link between energy use, government expenditure and financial development in Venezuela from 1971 to 2013. The study employed an autoregressive distributed lag (ARDL) model and found a positive link between energy use and environmental degradation. The study further revealed a negative relationship between land degradation and government expenditure. A study carried out by Uwazi [55] and You and Haung [56] examined the link between green growth and environmental sustainability in the OECD. The study looked at 30 provinces using panel data. The results show a positive relationship between government spending and green growth. A similar study was done by Oyebanji et al. [57] who conducted a study on green growth and environmental sustainability in Nigeria. The study considered energy depletion, forestry, carbon dioxide and employed the ARDL model. The study found a negative relationship between carbon emission, environmental depletion, and greenhouse energy. On the other hand, a positive relationship was found between green growth and deforestation.

From the empirical literature reviewed, there is no consensus on how taxes influence emission. Certain authors support a positive relationship between the variable, others see no link, while others support a negative relationship. Given such a scenario, the study, therefore, contributes to the existing literature by examining the influence of tax on emission.

3. Data and research methodology

This paper is based on associations among tax structures, environmental variables, income, production, transport, eco-innovation and green investments in a panel of 28 economies over a period of 7 years, that is, 2010 to 2017. The 7-year period was deemed sufficient due to data availability and sufficient cross-sections. These variables were chosen as they have a potential impact of reducing or increasing greenhouse gas emissions and environmental sustainability. The generalised below equations form the basis of the hypothesis.

$$\begin{aligned} Log\ GHG_{it} = {} & \alpha_1 + \alpha_2 Log\ GHG_{it-1} + \alpha_3 LogETT_{it} + \alpha_4 LogGDP_{it} + \alpha_5 LogPDN_{it} \\ & + \alpha_6 LogECO_{it} + \alpha_7 LogEC_{it} + \alpha_8 LogGRD_{it} + \alpha_9 LogGE_{it} \\ & + \varepsilon_{it} \ldots \ldots \ldots \ldots \ldots \ldots \end{aligned} \tag{1}$$

$$\begin{aligned} Log\ ANS_{it} = {} & \alpha_1 + \alpha_2 Log\ ANS_{it-1} + \alpha_3 LogETT_{it} + \alpha_4 LogGDP_{it} + \alpha_5 LogPDN_{it} \\ & + \alpha_6 LogECO_{it} + \alpha_7 LogEC_{it} + \alpha_8 LogGRD_{it} + \alpha_9 LogGE_{it} \\ & + \varepsilon_{it} \ldots \ldots \ldots \ldots \ldots \ldots \end{aligned} \tag{2}$$

And more specifically,

$$\begin{aligned} Log\ GHG_{it} = {} & \alpha_1 + \alpha_2 Log\ GHG_{it-1} + \alpha_3 LogENT_{it} + \alpha_4 LogTRT_{it} + \alpha_5 LogGDP_{it} \\ & + \alpha_6 LogPDN_{it} + \alpha_7 LogECO_{it} + \alpha_8 LogEC_{it} + \alpha_9 LogGRD_{it} + \alpha_{10} LogGE_{it} \\ & + \varepsilon_{it} \ldots \end{aligned} \tag{3}$$

And also,

$$Log\ ANS_{it} = \alpha_1 + \alpha_2 Log\ ANS_{it-1} + \alpha_3 LogENT_{it} + \alpha_4 LogTRT_{it} + \alpha_5 LogGDP_{it} + \alpha_6 LogPDN_{it} + \alpha_7 LogECO_{it} + \alpha_8 LogEC_{it} + \alpha_9 LogGRD_{it} + \alpha_{10} LogGE_{it} + \varepsilon_{it} \dots \quad (4)$$

Where,

*Log*GHG shows greenhouse gas emissions. $Log\text{GHG}_{it-1}$ is the lagged dependent variable of greenhouse gas emissions. *Log*ANS illustrates is an indicator of environmental sustainability.$Log\text{ANS}_{it-1}$ is the lagged dependent variable of environmental sustainability. *Log*ECO indicates the Eco-innovation index with a point system eco-innovation indicator. *Log*PDN indicates production scores. *Log*GRD illustrates green research and development. *Log*GE demonstrates the government expenditure. *Log*ENT is Energy Tax. *Log*TRT is Transport Tax. *Log*EC is energy consumption. *Log*GDP represents income and/or economic growth. *Log*ETT is Environmental Tax. In this regard, the table below outlines the variables employed in this study and their sources.

An environmental tax is a certain amount that is imposed to environment polluters [21]. For the purposes of this study, environment tax includes the energy tax and transport tax and it is expected to reduce greenhouse gas emissions and increase environment sustainability depending with the price elasticity demand of energy and transport. The rationale is that environmental tax should create awareness to switch to energy efficiency and turn to other clean alternative fuels. Energy tax is the tax that is levied on the energy sector for polluting the environment [58]. Energy tax includes the consumption of petrol, diesel, petrol, diesel, biofuels, electricity consumption and carbon fuels [21]. Transport tax is a tax that pertains to the use of all vehicles in the European Union [23]. The aforementioned taxes are expected to reduce the green house emission at the same time promote environmental sustainability. Energy consumption is the energy used in both industries and at household level which is measured in tonnes of oil [59]. The study expects energy consumption to increase greenhouse gas emissions and reduce environmental sustainability.

Green Research and Development is defined as new innovations introduced to minimise emissions and climate change in the European Union [60]. A positive relationship between Green Research and Development and environment sustainability is expected while an inverse relationship on greenhouse gas emission is expected. The rationale is that new innovations provide better ways of energy use that minimises climate change. Likewise, eco-innovation includes all ideas from stakeholders to develop new products and processes that reduces environmental degradation [61]. Eco-innovation reduces the greenhouse emissions and increases the environmental sustainability.

Production is a scientific procedure of turning all the inputs into consumable goods and services of a certain good and service [62]. Since production makes use of energy, the variable is expected to positively contribute to greenhouse gas emission and reduces the environmental sustainability in the European Union. Gross Domestic Product entails the value of all goods and services produced in the European Union countries over a specified period [63]. The priori expectation is that GDP increases greenhouse emission and decreases environment sustainability in the short-run while betters environment sustainability in the long-run. Government spending is the money spent by the government in acquiring public goods and services [64]. Government expenditure is expected to increase greenhouse expenditure if less or no expenditure is done on reducing climate change. On the other hand, the greenhouse expenditure is likely to reduce if the government spent much on improving climate change.

From **Table 1**, the logarithm of greenhouse gas along with logarithm of adjusted net savings (excluding particulate emission damage) depicts the dependent variables. The remaining variables are all explanatory variables. All the variables were extracted from the Eurostat database with the exception of the logarithm of adjusted net savings (excluding particulate emission damage) which is the only variable gathered from the World Development Indicators (World Bank) database.

3.1 Estimation technique

The paper deployed the panel dynamic Generalised Method of Moments (GMM) as the main approach to address problems connected with, heteroskedasticity serial correlation and heterogeneity [65]. The GMM captures several common estimators that offers a valuable basis for comparison purposes [66]. It is considered as one of the best methods since it not biased, consistent compared to Fixed effects, Pooled Estimates and Ordinary Least Squares [67]. Furthermore, the model allows researchers to make use of many independent variables without facing the endogeneity problems. Thus, the model provides the robust coefficients through the automatic correction of biasness. Several researchers such as Leve and Kapingura [68], Meraya et al. [69] and Nayan et al. [70] have employed the GMM.

Variable	Definition	Unit	Source
*Log*ETT	Logarithm of Environmental Tax	Percentage of gross domestic product (GDP)	Eurostat
*Log*ENT	Logarithm of Energy Tax	Percentage of gross domestic product (GDP)	Eurostat
*Log*TRT	Logarithm of Transport Tax	Percentage of gross domestic product (GDP)	Eurostat
*Log*GHG	Logarithm of Greenhouse gas emissions	Greenhouse gas emissions per capita	Eurostat
*Log*EC	Logarithm of Energy consumption	Thousand tonnes of oil equivalent	Eurostat
*Log*GRD	Logarithm of Green Research & Development	Percentage of gross domestic product (GDP)	Eurostat
*Log*GDP	Logarithm of GDP	Current prices, million units of national currency	Eurostat
*Log*GE	Logarithm of Government Expenditure	Percentage of gross domestic product (GDP)	Eurostat
*Log*ECO	Logarithm of Eco-innovation Index	Yearly scores against the EU = 100 averaged score.	Eurostat
*Log*PDN	Logarithm of production	Yearly scores based on Index, 2015 = 100	Eurostat
*Log*ANS	Logarithm of adjusted net savings, excluding particulate emission damage	Current US$	World Development Indicators (World Bank)

Note: The Logarithm of Greenhouse gas emissions and Logarithm of adjusted net savings, excluding particulate emission damage indicates the dependent variable. The remaining variables are all explanatory variables.

Table 1.
Showing detailed description of variables.

For the purposes of this study, it is apparent that Eq. (1) is comprised of country time effects as well as country fixed effects which is inevitably generates the problem of unobserved country-specific heterogeneity. Thus, Arellano and Bond [71] highlights that GMM is able to transform such particular equations through first difference estimators. Research also shows that the GMM approach is largely suitable in surveys where the cross-section identifiers are greater in quantity (in this study, $N = 28$) against small time periods (in this article, $T = 7$ years) [72]. Overall, panel regression problems such as heterogeneity, heteroskedasticity along with serial correlation can be significantly addressed by using a panel GMM technique [73, 74].

4. Empirical analysis and results

This part of the survey outlines the findings of the study which includes the descriptive statistics, panel unit root test and the GMM results. The following section discusses the descriptive statistics.

4.1 Descriptive statistics

Table 2 illustrates a detailed view of the statistical characteristics of the variables used in this study. It is apparent that the mean of the considered variables is located between their own minimum and maximum values. As well, most of these factors are negatively skewed (63.6%) but only 36.4% demonstrates positive skewness. In this case, transport tax, energy consumption, green research and development, government expenditure, eco-innovation index, production, and environmental sustainability are negatively skewed. On the other hand, environmental tax, energy tax, greenhouse gas emissions, and economic growth are positively skewed. The positive values generated through kurtosis imply that all the variables have leptokurtic attributes.

Variable	Min.	Max.	Mean	Std. Dev.	Skewness	Kurtosis	Observation
*Log*ETT	0.196	0.617	0.405	0.102	0.114	2.254	224
*Log*ENT	0.033	0.521	0.286	0.103	0.246	2.575	224
*Log*TRT	−1.301	0.190	−0.393	0.353	−0.670	2.867	224
*Log*GHG	0.699	1.423	0.959	0.143	0.637	3.319	224
*Log*EC	2.589	5.315	4.206	0.620	−0.323	2.902	224
*Log*GRD	−0.420	0.572	0.128	0.256	−0.157	2.009	224
*Log*GDP	3.820	7.584	5.517	0.868	0.074	2.325	224
*Log*GE	1.420	1.814	1.654	0.066	−0.389	3.2358	224
*Log*ECO	1.301	2.173	1.915	0.1728	−0.709	3.046	224
*Log*PDN	1.203	2.124	1.926	1.926	−4.948	25.678	224
*Log*ANS	0	11.7289	8.969	3.535	−2.040	5.464	224

Source: Authors compilation.

Table 2.
Statistical summary of variables.

4.2 Panel unit root test analysis

Table 3 shows that when the Augmented Dickey-Fuller test (ADF) tests; Levin, Lin, and Chu (LLC) and the Im-Pesaran-Shin (IPS) were employed the time series is not affected by the presence of unit roots. As such, through deploying a null hypothesis that a specific time series is non-stationary all the variables demonstrates that they are stationary at the first-order differenced series for all ADF, LLC and IPS tests (at 1% significant level) employing the first-generation panel unit-roots. Although variables such as transport tax, production, and environmental sustainability were not confirmed using the IPS test the other two remaining tests argument in favor of the general findings of the paper.

Using logarithm of greenhouse gas emissions as the dependent variable and total environmental tax as the main independent variable **Table 4** outlines the results of the research about the Pooled Ordinary Least Square (OLS) model, Fixed Effect (FE) and Random Effect (RE) models widely regarded as the static models. The details in **Table 4** demonstrates that Hausman test produces a chi-square value of 270.32 which is also significant at 5% ($p = 0.000 < 0.05$). This shows that we will reject the null hypothesis which illustrates that the RE model is suitable in favor of the alternative hypothesis which explains that the FE model is suitable. The dynamic nature of the FE model is further analysed using the two-step GMM model.

	At Level			At 1st Difference		
Variable	**ADF statistic**	**LLC Statistic**	**IPS Statistic**	**ADF statistic**	**LLC Statistic**	**IPS Statistic**
*Log*ETT	1.963 (0.025)**	−7.686 (0.000)***	0.886 (0.8122)	9.339 (0.000)***	−13.951 (0.000)***	−3.378 (0.000)***
*Log*ENT	2.467 (0.007)***	−8.548 (0.000)***	1.483 (0.931)	9.068 (0.000)***	−10.568 (0.000)***	−3.431 (0.000)***
*Log*TRT	2.861 (0.002)***	−8.234 (0.000)***	—	19.540 (0.000)***	−23.409 (0.000)***	—
*Log*GHG	8.940 (0.000)***	−8.525 (0.000)***	−1.817 (0.035)**	5.490 (0.000)***	−9.777 (0.000)***	−2.454 (0.007)***
*Log*EC	7.125 (0.000)***	−11.099 (0.000)***	−1.622 (0.052)*	9.722 (0.000)***	−19.600 (0.000)***	−2.532 (0.006)***
*Log*GRD	4.486 (0.000)***	−6.545 (0.000)***	−0.060 (0.476)	19.727 (0.000)***	−10.522 (0.000)***	−3.990 (0.000)***
*Log*GDP	−1.634 (0.949)	9.158 (1.000)	10.826 (1.000)	2.395 (0.008)***	−16.523 (0.000)***	−0.606 (0.272)
*Log*GE	7.899 (0.000)***	−3.947 (0.000)***	1.734 (0.9586)	10.764 (0.000)***	−15.177 (0.000)***	−3.398 (0.000)***
*Log*ECO	3.003 (0.001)***	−3.321 (0.004)***	−1.948 (0.026)**	23.576 (0.000)***	−13.886 (0.001)***	−4.826 (0.000)***
*Log*PDN	−2.484 (0.994)	5.508 (1.0000)	—	10.630 (0.000)***	−9.284 (0.000)***	—
*Log*ANS	9.972 (0.0000)***	−2.517 (0.006)***	—	18.614 (0.000)***	−20.013 (0.000)***	—

*Notes: ***;**;* mean significant at 1%, 5%, 10% level of significance respectively. Numbers in brackets are* p-values.

Table 3.
Showing the panel unit root test results.

	Pooled Model		Random Effect Model		Fixed Effect Model	
	Coefficient	Standard Error	Coefficient	Standard Error	Coefficient	Standard Error
*Log*ETT	0.018 (0.851)	0.093	−0.032 (0.599)	0.062	−0.094 (0.078)*	0.053
*Log*EC	0.028 (0.241)	0.024	0.417 (0.000)***	0.060	0.563 (0.000)***	0.086
*Log*GRD	0.298 (0.000)***	0.056	−0.028 0.442	0.037	−0.076 (0.017)**	0.032
*Log*GDP	−0.065 (0.000)***	0.016	−0.336 (0.000)***	0.041	−0.707 (0.000)***	0.054
*Log*GE	−0.367 (0.041)**	0.178	−0.046 0.489	0.066	−0.224 (0.000)***	0.060
*Log*ECO	−0.031 (0.667)	0.074	0.026 0.399	0.030	−0.025 (0.339)	0.026
*Log*PDN	−0.062 (0.011)**	0.024	0.194 (0.001)***	0.060	0.678 (0.000)***	0.078
Constant	1.942 (0.0000)***	0.292	0.729 (0.005)***	0.260	1.654 (0.000)***	0.449
R^2	0.239		0.012		0.003	
Wald (χ^2)			86.07			
F statistic	9.70				34.14	
Breusch-Pagan test (χ^2)			654.5 (0.000)***			
Hausman test (χ^2)					270.3 (0.000)***	
No. of observations	224		224		224	

*Notes: ***; **; * mean significant at 1%, 5% and 10% significance level, respectively. Numbers in brackets are* p*-values.*

Table 4.
Estimates of static panel data for total environmental tax: Case of Greenhouse Emissions.

The results found in **Table 4** are also generally congruent with outcomes found in **Table 5**. For example, the Hausman test generates a chi-square estimate of 15.24 which is also significant at 5% ($p = 0.0330 < 0.05$) supports the FE model which permits the study to use the two-step GMM analysis procedure.

The findings generated in **Tables 4** and **5** are also generally confirmed with results in **Table 6** (although in this case energy tax and transport tax are the main independent variables). For instance, the Hausman test shows a chi-square estimate of 195.27 which is also significant at 5% ($p = 0.000 < 0.05$) favoring the FE model. As such, the two-step GMM analytical process will be applied.

In **Table 7**, the Hausman test generates a chi-square estimate of 18.41 which is also significant at 5% ($p = 0.0184 < 0.05$) thereby supporting the FE model.

Table 8 presents the outcomes acquired by running the two-step GMM analytical method within the short-run context with regards to total environmental tax as the main independent variable. We begin first by evaluating greenhouse gas emissions as the dependent variable. To begin, the lagged factor $Log\text{GHG}_{it-1}$ of greenhouse gas emissions indicates a positive and significant relationship with greenhouse gas emissions. Hence, a 1% increase in lagged greenhouse gas emissions

	Pooled Model		Random Effect Model		Fixed Effect Model	
	Coefficient	**Standard Error**	**Coefficient**	**Standard Error**	**Coefficient**	**Standard Error**
*Log*ETT	3.719 (0.076)*	2.088 (0.230)	3.199	2.666	1.963 (0.516)	3.017
*Log*EC	1.889 (0.000)***	0.531 (0.385)	1.230	1.414	−11.462 (0.021)**	4.917
*Log*GRD	4.266 (0.001)***	1.240 (0.881)	−0.227	1.517	−3.132 (0.087)*	1.818
*Log*GDP	0.893 (0.013)**	0.357 (0.127)	1.478	0.968	−1.589 (0.606)	3.077
*Log*GE	−13.285 (0.001)***	3.977 (0.018)**	−6.706	2.824	−6.407 (0.061)***	3.401
*Log*ECO	0.200 (0.899)	1.653 (0.072)*	2.400	1.330	0.981 (0.507)	1.477
*Log*PDN	−0.582 (0.278)	0.535 (0.948)	0.091	1.398	10.747 (0.016)**	4.403
Constant	16.742 (0.011)**	6.513 (0.942)	0.710	7.025	53.571 (0.038)**	25.632
R^2	0.382		0.324		0.301	
Wald (χ^2)			26.43			
F statistic	19.07				3.69	
Breusch-Pagan test (χ^2)			489.70 (0.000)***			
Hausman test (χ^2)					15.24 (0.033)**	
No. of observations	224		224		224	

*Notes: ***; **; * mean significant at 1%, 5% and 10% significance level, respectively. Numbers in brackets are p-values.*

Table 5.
Estimates of static panel data for total environmental tax: Case of Environmental Sustainability.

triggers a 0.218% increase in greenhouse gas emissions. This diagnosis implies that when past greenhouse gas emissions in the EU economies rise then they will also stimulate current emission levels. This is confirmed by continued increase in emissions globally [60, 75] which require to be lowered.

Secondly, the total environmental tax shows a positive and highly significant association with greenhouse gas emissions. In this context, a single rise in total environmental tax leads to a 0.22 increase in greenhouse gas emissions. However, this study finding conflicts with [76] who noticed significantly small and even negative carbon leakage after unilateral environmental tax reforms were integrated in Europe between the studied periods 1995 to 2005. Third, a 1% increase in energy consumption also results in a significant 0.73% rise in emissions thereby agreeing with [77] analysis on 116 countries over the period 1990 to 2014. Fourth, a percentage rise in green research and development in the short-run is also leading to a 0.36% increase in greenhouse gas emissions. However, this finding contradicts Fernández, López and Blanco's [78] survey on 15 European Union countries, the United States and China between 1990 and 2013 and spotlights that green research and development adds positively to a decline in emissions in developed countries.

	Pooled Model		Random Effect Model		Fixed Effect Model	
	Coefficient	**Standard Error**	**Coefficient**	**Standard Error**	**Coefficient**	**Standard Error**
*Log*ENT	0.109 (0.215)	0.087	−0.056 (0.293)	0.054	−0.067 (0.148)	0.046
*Log*TRT	−0.100 (0.001)***	0.029	0.032 (0.419)	0.039	−0.015 (0.685)	0.036
*Log*EC	0.015 (0.498)	0.023	0.411 (0.000)***	0.060	0.564 (0.000)***	0.087
*Log*GRD	0.245 (0.000)***	0.056	−0.029 (0.433)	0.037	−0.075 (0.022)***	0.032
*Log*GDP	−0.060 (0.000)***	0.016	−0.334 (0.000)***	0.040	−0.703 (0.000)***	0.054
*Log*GE	−0.060 (0.752)	0.190	−0.045 (0.501)	0.066	−0.226 (0.000)***	0.060
*Log*ECO	0.058 (0.447)	0.076	0.022 (0.467)	0.031	−0.025 (0.347)	0.026
*Log*PDN	−0.092 (0.000)***	0.025	0.191 (0.001)***	0.059	0.672 (0.000)***	0.078
Constant	1.290 (0.000)***	0.337	0.768 (0.003)***	0.262	1.618 (0.000)***	0.451
R^2	0.284		0.010		0.003	
Wald (χ^2)			87.09			
F statistic	10.68				29.53	
Breusch-Pagan test (χ^2)			659.64 (0.000)***			
Hausman test (χ^2)					195.27 (0.000)***	
No. of observations	224				224	

*Notes: ***; **; * mean significant at 1%, 5% and 10% significance level, respectively. Numbers in brackets are* p-*values.*

Table 6.
Estimates of static panel data for total energy tax and transport tax: Case of Greenhouse-gas Emissions.

However, other remaining variables indicates negative and significant links to greenhouse gas emissions. For example, a percentage increase in economic growth leads to a 0.40% significant decrease in emissions. Nonetheless, this study outcomes disagrees with Salahuddin et al. [79] research on Kuwait for the period 1980–2013 by applying the autoregressive distributed lag (ARDL) bounds testing approach and adds that economic growth motivates emissions in both short-run and long-run. In another context, a 1% rise in government expenditure significantly lowers greenhouse gas emissions by 0.815%. However, [80] studied the Venezuelan context over the period from 1971 to 2013 and contributes that government expenditure has a positive effect on environmental degradation-emissions. In addition, the eco-innovation rating is also responsible for decreasing emissions significantly by 0.0039% in these studied EU countries in the short-run. Using the GMM technique on China's 30 provinces during 2000–2013, [81] also contributes that environmental innovation resources along with green knowledge innovation are essential components for emissions reduction. The results of this study also demonstrates that a

	Pooled Model		Random Effect Model		Fixed Effect Model	
	Coefficient	**Standard Error**	**Coefficient**	**Standard Error**	**Coefficient**	**Standard Error**
*Log*ENT	4.516 (0.025)**	1.999	0.839 (0.717)	2.318	−1.207 (0.639)	2.571
*Log*TRT	0.382 (0.564)	0.662	2.106 (0.113)	1.327	5.084 (0.012)**	2.008
*Log*EC	1.787 (0.001)***	0.522	1.182 (0.412)	1.44	−12.918 (0.009)***	4.896
*Log*GRD	4.484 (0.001)***	1.286	−0.362 (0.813)	1.525	−3.436 (0.059)**	1.805
*Log*GDP	0.902 (0.012)**	0.3572	1.397 (0.159)	0.992	−0.837 (0.784)	3.047
*Log*GE	−13.352 (0.002)***	4.334	−6.780 (0.016)**	2.827	−6.056 (0.073)*	3.357
*Log*ECO	0.422 (0.809)	1.742	2.122 (0.116)	1.352	0.578 (0.696)	1.475
*Log*PDN	−0.737 (0.197)	0.569	0.370 (0.786)	1.438	9.717 (0.027)**	4.355
Constant	17.462 (0.024)**	7.699	3.327 (0.647)	7.264	60.900 (0.017)**	25.302
R^2	0.3883		0.274		0.304	
Wald (χ^2)			27.19			
F statistic	17.06				4.06	
Breusch-Pagan test (χ^2)			482.89 (0.000)***			
Hausman test (χ^2)					18.41 (0.0184)**	
No. of observations	224				224	

*Notes: ***; **; * mean significant at 1%, 5% and 10% significance level, respectively. Numbers in brackets are p-values.*

Table 7.
Estimates of static panel data for total energy tax and transport tax: Case of Sustainability.

1% increase in production will also likely reduce emissions by a significant 0.40%. However, Ganda [82] survey on the BRICS (Brazil, Russia, India, China, and South Africa) using panel data from 1992 to 2014 express that production practice, through industrial initiative adds to emissions.

The second part of this section will examine the short-run results by examining environmental sustainability as the dependent variable. In this case, a 1% rise in past adjusted net savings, excluding particulate emission damage, which is the proxy for environmental sustainability ($Log\text{ANS}_{it-1}$) will significantly improve current environmental sustainability levels in the scrutinised EU economies by 0.235%. Secondly, a percentage increase in total environmental tax leads to a 2.88% increase in environmental sustainability. Kosonen [83] suggests that environmental taxes are major instruments that governments can deploy in order to achieve sustainability although their regressive effects require extensive consideration. Thirdly, a 1% increase in energy consumption significantly rises environmental sustainability by 5.56%. This outcome is elaborated by [84] study on OECD

	*Log*GHG		*Log*ANS	
	Coefficient	**Standard Error**	**Coefficient**	**Standard Error**
Log GHG_{it-1}	0.218 (0.004)***	0.075		
Log ANS_{it-1}			0.235 (0.000)***	0.009
*Log*ETT	0.222 (0.006)***	0.0803589	2.877 (0.000)***	0.681
*Log*EC	0.731 (0.000)***	0.1221115	5.556 (0.000)***	1.159
*Log*GRD	0.362 (0.000)***	0.084	−0.753 (0.053)*	0.389
*Log*GDP	−0.403 (0.000)***	0.083	−1.324 (0.094)*	0.790
*Log*GE	−0.815 (0.000)***	0.184	−10.793 (0.000)***	0.877
*Log*ECO	−0.004 (0.091)*	0.038	−1.542 (0.000)***	0.109
*Log*PDN	−0.396 (0.000)***	0.131	2.910 (0.000)***	0.681
Constant	1.862 (0.000)***	0.377	5.028 (0.000)***	2.380
Wald (χ^2)	88.56 (0.000)		11340.75 (0.000)	
Arellano-Bond test for AR (1) in first differences	z = −0.97 Pr > z = 0.003		z = −1.15 Pr > z = 0.025	
Arellano-Bond test for AR (2) in first differences	z = −0.52 Pr > z = 0.600		z = 0.29 Pr > z = 0.769	
Hansen test of overidentifying. Restrictions	Chi-square = 30.79 Prob > chi2 = 0.683		Chi-square = 21.06 Prob > chi2 = 1.000	
Sargan test of overidentifying. Restrictions	Chi-square = 16.49 Prob > chi2 = 0.284		Chi-square = 106.77 Prob > chi2 = 0.2	
No. of observations	196		196	

*Notes: ***; **; * mean significant at 1%, 5% and 10% significance level, respectively. Numbers in brackets are* p-values.

Table 8.
Findings of GMM short-run results as the dynamic regression approaches: In case of total environmental tax.

economies from 1980 to 2011 using the STIRPAT model and highlights that renewable energy consumption promotes sustainability by lowering emissions while non-renewable energy use increase emissions thereby destroying the natural environment.

Fourth, when short-run green research and development increased by 1% then environmental sustainability will significantly decrease by 0.75%. This finding conflicts with [85] study on US electric generators who adds that short-run decisions to integrate green technologies also provide significant emission reduction opportunities even before new technologies have been fully integrated on a broadened scale.

Furthermore, income is also found to be lowering environmental sustainability in the short-run for the studied EU countries. In this context, a percentage increase in economic development significantly decreases sustainability by 1.32%. Nevertheless, Ganda [86] study on OECD economies also highlights that disagrees with these outcomes as income is ascertained to increase environmental sustainability by 17.8% in the short-run. Another variable, government expenditure is also ascertained to significantly lower environmental sustainability by 10.79% when it increases by a single percent. Then, a 1% rise in eco-innovation is also accountable to a significant decrease estimated at 1.54% of environmental sustainability. However, a 1% increase in production will significantly heighten environmental sustainability by 2.91%. As such, Severo, de Guimarães, Dorion and Nodari [87] having explored the Brazilian Metal-Mechanic industry posits that cleaner production positively influences environmental sustainability.

Table 9 also depicts the results obtained by implementing a two-step GMM analysis process. The presentation disaggregates total environmental tax by identifying energy tax and transport tax as the main independent factors in this analysis. As previously done in the previous section, we commence by initially assessing greenhouse gas emissions as the dependent variable. In this context, it is evident that the lagged variable factor $Log\text{GHG}_{it-1}$ of greenhouse gas emissions indicates a positive and significant link with greenhouse gas emissions. More precisely, a percentage increase in lagged greenhouse gas emissions stimulates a 0.29% rise in greenhouse gas emissions. These results concur with earlier results determined in this paper when total environmental tax was analysed as an aggregate green tax proxy.

In addition, energy tax demonstrates a positive and significant connection with greenhouse gas emissions. As such, a 1% rise in energy tax is sufficient to increase emissions by 0.10%. However, Solaymani [88] study on Malaysia found out that energy tax can reduce emissions although carbon tax was found to be a more effective tax instrument for emissions reduction programs. Furthermore, the paper outcomes show that transport tax shows a negative and highly significant association with greenhouse gas emissions. In this context, a single rise in transport tax leads to a 0.13% decrease in greenhouse gas emissions. González and Hosoda [89] also conducted a study in Japan between 2004 and 2013 using the Bayesian structural time series model and they highlight that the integration of fuel tax has unequivocally minimised aircraft emissions.

As well, the research illustrates that a 1% increase in energy consumption also results in a significant 0.47% rise in emissions thereby agreeing with findings presented in **Table 8**. Conversely, the results in **Table 9** further indicates that a percentage rise in green research and development generates a 0.15% decrease in greenhouse gas emissions thereby supporting [79] study on 15 European Union countries. Furthermore, economic growth has a negative and significant association with emissions. As such, a 1% increase in income stimulates a 0.28% reduction in emissions. However, Magazzino [90] study on Italy over the period 1970 to 2006 demonstrates a bidirectional causality link between economic growth and emissions.

The other outstanding variables indicate positive and significant links to greenhouse gas emissions. For instance, a percentage increase in government expenditure leads to a 0.176% significant rise in emissions. Contradicting with these findings [91] study on a panelised data of 94 countries between 1970 and 2008 illustrates that government expenditure exercise a significant direct influence in reducing the amount of emissions. As well, the eco-innovation rating is also responsible for rising emissions significantly by 0.11%. However, Costantini et al. [92] exploration of European industries confirm that both indirect and direct impacts of eco-innovations assist lessening environmental degradation although the strength varied throughout the industry value chain. The outcomes of the research also confirm

	*Log*GHG		*Log*ANS	
	Coefficient	**Standard Error**	**Coefficient**	**Standard Error**
$Log\ GHG_{it-1}$	0.286 (0.000)***	0.062		
$Log\ ANS_{it-1}$			0.163 (0.000)* **	0.019
*Log*ENT	0.100 (0.066)*	0.055	−4.369 (0.011)**	1.719
*Log*TRT	−0.130 (0.001)***	0.038	5.740 (0.000)***	1.397
*Log*EC	0.466 (0.000)***	0.059	0.539 (0.644)	1.166
*Log*GRD	−0.149 (0.001)***	0.043	1.750 (0.099)*	1.062
*Log*GDP	−0.280 (0.000)***	0.050	1.304 (0.056)*	0.684
*Log*GE	0.176 (0.008)***	0.066	−12.552 (0.000)***	3.015
*Log*ECO	0.113 (0.000)***	0.019	0.341 (0.008)***	0.128
*Log*PDN	0.155 (0.012)***	0.062	−5.987 (0.001)***	1.848
Constant	−0.615 (0.010)***	0.244	33.271 (0.000)***	8.032
Wald (χ^2)	411.85 (0.000)		3202.15 (0.000)	
Arellano-Bond test for AR (1) in first differences	z = −2.09 Pr > z = 0.036		z = −1.10 Pr > z = 0.022	
Arellano-Bond test for AR (2) in first differences	z = 0.32 Pr > z = 0.752		z = 0.60 Pr > z = 0.547	
Hansen test of overidentifying. Restrictions	Chi-square = 19.82 Prob > chi2 = 0.898		Chi-square = 20.10 Prob > chi2 = 0.890	
Sargan test of overidentifying. Restrictions	Chi-square = 43.9 Prob > chi2 = 0.38		Chi-square = 38.51 Prob > chi2 = 0.111	
No. of observations	196			

*Notes: ***; **; * mean significant at 1%, 5% and 10% significance level, respectively. Numbers in brackets are* p-values.

Table 9.
Findings of GMM short-run results as the dynamic regression approaches: In case of energy tax and transport tax.

that as production in the short-run increases by 1% emissions also heightens by 0.15%. Likewise, Phalan et al.'s [93] survey on the Brazilian beef industry expresses that production is highly unlikely to help lower emissions, and is possibly likely to exacerbate deforestation.

The remaining segment of this section will evaluate the GMM findings through scrutinising environmental sustainability as the dependent variable. Thus, from **Table 9**, if lagged environmental sustainability (*Log*ANS_{it-1}) increases by 1% then a

0.16% improvement in current environmental sustainability levels in the studied EU countries is apparent thereby supporting outcomes validated in **Table 8**. Secondly, a percentage increase in total energy tax leads to a 4.37% decrease in environmental sustainability. Likewise, Choi et al.'s [94] survey on the United States gas taxes and fuel subsidy policy explains that in situations where part of gasoline tax revenue is prioritised towards subsidising biofuel production then better resource consumption and mitigated emissions will be evidenced. However, this paper outcomes indicate that a 1% rise in transport tax increase environmental sustainability by 5.74%.

The paper results also demonstrate that a 1% increase in energy consumption significantly rises environmental sustainability by 0.54%. Furthermore, it can be ascertained that if green research and development increased by 1% then environmental sustainability will significantly increase by 1.75%. Moreover, a percentage rise in income motivates a 1.30% rise in environmental sustainability. Hatfield-Dodds et al. [95] study on Australia also contributes that it is quite difficult to decouple economic growth and environmental outcomes and mobilisation of technologies and engagement of environmental incentives are essential for advancement towards sustainable prosperity. The research outcomes also show that a 1% increase in government expenditure is also ascertained to significantly lower environmental sustainability by 12.6%. In addition, a 1% rise in eco-innovation is also accountable to a significant increase estimated at 0.34% of environmental sustainability. Nevertheless, a 1% increase in production will significantly lessen environmental sustainability by 5.99%.

Table 8 which was presented earlier in this section outline the regression findings in the short-run scenario in case where environmental tax was identified as the main independent variable. **Table 10** above extends the discussion by examining the association involving environmental tax as the primary independent factor to both emissions and environmental sustainability but on a long-run setting. In detail, it is evident that environmental tax form a positive relationship with both greenhouse gas emissions and environmental sustainability (although it is significant in this context). Likewise, energy consumption shows a significantly positive link with

	*Log*GHG		*Log*ANS	
	Coefficient	**Standard Error**	**Coefficient**	**Standard Error**
*Log*ETT	0.004 (0.970)	0.102	2.642 (0.000)***	0.674
*Log*EC	0.513 (0.000)***	0.145	5.320 (0.000)***	1.157
*Log*GRD	0.144 (0.241)	0.123	−0.988 (0.012)**	0.394
*Log*GDP	−0.622 (0.000)***	0.110	−1.560 (0.049)**	0.792
*Log*GE	−1.034 (0.000)***	0.209	−11.028 (0.000)***	0.881
*Log*ECO	−0.222 (0.025)**	0.099	−1.778 (0.000)***	0.108
*Log*PDN	−0.614 (0.000)***	0.164	2.675 (0.000)***	0.685

*Notes: ***; **; * mean significant at 1%, 5% and 10% significance level, respectively. Numbers in brackets are p-values.*

Table 10.
Findings of GMM long-run results as the dynamic regression approach: in case of total environmental tax.

	*Log*GHG		*Log*ANS	
	Coefficient	**Standard Error**	**Coefficient**	**Standard Error**
*Log*ENT	−0.185 (0.021)**	0.080	−4.532 (0.008)***	1.713
*Log*TRT	−0.416 (0.000)***	0.074	5.577 (0.000)***	1.400
*Log*EC	0.180 (0.058)*	0.095	0.376 (0.745)	1.157
*Log*GRD	−0.435 (0.000)***	0.078	1.587 (0.136)*	1.064
*Log*GDP	−0.566 (0.000)***	0.058	1.142 (0.099)*	0.692
*Log*GE	−0.110 (0.292)*	0.104	−12.714 (0.000)***	3.005
*Log*ECO	−0.173 (0.005)***	0.061	0.178 (0.156)	0.126
*Log*PDN	−0.131 (0.193)*	0.101	−6.150 (0.000)***	1.842

*Notes: ***; **; * mean significant at 1%, 5% and 10% significance level, respectively. Numbers in brackets are* p-*values.*

Table 11.
Findings of GMM long-run results as the dynamic regression approach: In case of energy tax and transport tax.

both emissions and environmental sustainability. Green research and development produce a positive link with emissions but its connection with environmental sustainability is significantly negative. The results further prove that economic growth, government expenditure, and eco-innovation show significant negative relationships to both emissions and environmental sustainability in the long-term. Lastly, production generates a significantly negative link with emissions but its association with environmental sustainability is significantly positive.

Table 9 of this part of the study produced short-run associations by disintegrating total environmental tax through isolating energy tax and transport tax as the main independent variables. **Table 11** expand this analysis by identifying the association of these explanatory variables against both emissions and environmental sustainability within a long-run basis. In brief, energy tax, government expenditure and production produces a significantly negative connection with both emissions and environmental sustainability. Other findings confirm that transport tax, green research and development, economic growth and eco-innovation demonstrate negative and positive associations with both emissions and environmental sustainability. The relationship involving energy use to both emissions and environmental sustainability is positive in both cases.

5. Discussion and implications

This section presents a detailed analysis of the study also highlights the implications of the research.

Table 12 provide useful insights about the context involving the association between total environmental tax and both greenhouse gas emissions along with environmental sustainability. Total environmental tax appears to be increasing emissions both on the short-and long-run scenario although it is found to be also

	Environmental tax				Energy tax and Transport tax			
	*Log*GHG		*Log*ANS		*Log*GHG		*Log*ANS	
	Short-Run	**Long-Run**	**Short-Run**	**Long-Run**	**Short-Run**	**Long-Run**	**Short-Run**	**Long-Run**
*Log*ETT	+	+	+	+				
*Log*EC	+	+	+	+	+	+	+	+
*Log*GRD	+	+	—	—	—	—	+	+
*Log*GDP	—	—	—	—	—	—	+	+
*Log*GE	—	—	—	—	+	—	—	—
*Log*ECO	—	—	—	—	+	—	+	+
*Log*PDN	—	—	+	—	+	—	—	—
*Log*ENT					+	—	—	—
*Log*TRT					—	—	+	+

Table 12.
Summary of GMM short-and long-run results.

simultaneously increasing environmental sustainability. This implies that while overall natural environmental effect as a result of imposing environmental tax improves there is also a need for EU economies to introduce specific green taxes which directly focus on particular environmental indicators so that emission reduction is effectually achieved. Moreover, there is a need to transform or remove particular environmental taxes which are not effectively achieving zero-emission targets. As well, taxes can be modified by adding regulatory instruments so that they are aligned with natural environmental objectives and goals. It is also apparent that energy consumption has been increasing the level of emissions and environmental sustainability. In this case, EU economics should continue expanding the integration of renewable energy and oppose further consumption of fossil fuels. There is evidence of renewable energy use in EU economies [13, 60] in pursuit of lower emissions which can possibly explain the improved environmental sustainability context. However, there is also a need to upgrade energy systems of green energy technologies so that they do not add to heightening emissions.

Green research and development is found to be highly effective when environmental taxes are emphasising of particular environmental measures instead of adopting a holistic environmental tax policy. For instance, when environmental tax was disaggregated the tax tools used managed to motivate green research and development to lower emissions and simultaneously raise environmental sustainability. Economic growth is quite effective in lowering the level of greenhouse gases whether environmental tax is aggregated and/or disaggregated. Of note is that economic growth effectively improve environmental sustainability in the short and long-run when EU economies use specific environmental taxes when adopting a comprehensive environmental tax instrument.

On the one hand, government expenditure is very efficient in lowering emissions in the short and long-run but is also not able to promote environmental sustainability during these periods in case where aggregate environmental tax is employed. On the other hand, the situation is also predominantly noticeable when environmental tax has been disaggregated (energy tax and transport tax) except that it increases emissions in the short-run. This indication shows that government expenditure in EU economies needs to focus on an inclusive approach which supports all issues related to sustainability instead of putting much emphasis on emissions alone. In this case, government expenditure should also include environmental standards and regulations and measures which heighten environmental sustainability.

It is also observable that eco-innovation is capable of lowering emissions whether environmental tax is aggregated or disintegrated. However, in the case where environmental tax is not aggregated, that is, specific eco-innovation improves environmental sustainability but it worsens environmental sustainability in case of total environmental tax. This shows the importance of introducing specific eco-innovation regulatory standards that fits different parts of the production and ultimate distribution of manufactured goods and services.

Although production in EU economies manage to lower emissions in cases where environmental is aggregated and/or not is has not been able to improve environmental sustainability. In this case, while production has managed emissions reduction targets the impacts of this procedure on other natural environmental components require to be upgraded.

Lastly, it is apparent that energy tax has been lessening environmental sustainability but transport tax has been effective in creating required environmental sustainability scenarios. Both these taxes are also effective in the long-run in lowering emissions although energy tax is found to ineffective in lowering emissions in the short-run. It is evident that the transport tax appears to be a more effective instrument to meet environmental goals when compared to energy tax in EU economies. In this case, there is a need to revise energy policy and regulatory instruments that deal with energy in these countries so that such tools are harmonising with sustainability goals and objectives.

6. Conclusion

The first findings presented regression results when the aggregate environmental tax was employed. These outcomes show that total environmental tax, energy consumption, green research and development significantly heightened emissions in the short-run scenario. The results further demonstrate that in the short-term economic growth, government expenditure, eco-innovation rating and production scores significantly lowered emissions. The results also confirm that total environmental tax, energy consumption and production significantly increase environmental sustainability in the short-run. Conversely, green research and development, economic growth, government expenditure, and eco-innovation significantly lower environmental sustainability in the short-run. The long-run results demonstrate that environmental tax and energy consumption develop a positive relationship with both greenhouse gas emissions and environmental sustainability respectively. In addition, green research and development generates a positive connection with emissions although its link with environmental sustainability is significantly negative. Economic growth, government expenditure, and eco-innovation illustrates a significant negative relationships to both emissions and environmental sustainability in the long-term. In the long-run, production produces a significantly negative association with emissions but a significantly positive relationship with environmental sustainability.

The second part of the results section outlined regression when disaggregated environmental tax (energy tax and transport tax) was deployed. Thus, in the short-term, energy tax, energy consumption, government expenditure, eco-innovation rating, and production scores spur a significant rise in emissions. However, transport tax, green research and development, and income influence lessens emissions in the short-run. Furthermore, energy tax and production significantly reduce environmental sustainability in the short-term. Nonetheless, transport tax, energy consumption, green research and development, income and eco-innovation significantly increase environmental sustainability in the short-run. The long-run findings proves that energy tax, government expenditure and production produces a significantly negative relationship with both emissions and environmental sustainability. As well, transport tax, green research and development, economic growth along with eco-innovation produce negative and positive associations with both emissions and environmental sustainability. Lastly, energy use shows a significantly positive link to both emissions and environmental sustainability.

Author details

Fortune Ganda[1*] and Rufaro Garidzirai[2]

1 Department of Accounting, Walter Sisulu University, Butterworth, South Africa

2 Department of Management, Walter Sisulu University, Butterworth, South Africa

*Address all correspondence to: fganda@wsu.ac.za, fortune.ganda@gmail.com and fochi555@yahoo.com

References

[1] Haites E. Carbon taxes and greenhouse gas emissions trading systems: what have we learned? Climate Policy. 2018;**18**:955-966

[2] Heerden JH, Blignaut J, Bohlmann H, Cartwright A, Diederichs N, Mander M. The Economic and Environmental Effects of a Carbon Tax in South Africa: A Dynamic CGE Modelling Approach. South African Journal of Economic and Management Sciences. 2016;**19**: 135-140

[3] Lee JY, Leem JH, Kim HC. Effects of traffic-related air pollution on susceptibility to infantile bronchiolitis and childhood asthma: a cohort study in Korea. Journal of Asthma. 2018;**55**: 223-230. DOI: 10.1080/02770903.2017.1313270

[4] deB Richter Jr, D, Houghton, RA. Gross CO2 fluxes from land-use change: implications for reducing global emissions and increasing sinks. Carbon Management, 2011;1: 41–47.

[5] Bartolini M, Bottani E, Eric H. Green warehousing: Systematic literature review and bibliometric analysis. Journal Clean Production. 2019;**226**: 242-258

[6] Metcalf GE. Paying for Pollution: Why a Carbon Tax is Good for America. New York: Oxford University Press; 2019. DOI: 10.1093/oso/9780190694197.001.0001

[7] Trinomics. Study on energy prices, costs and subsidies and their impact on industry and households. Rotterdam: Trinomics publications Europa.

[8] Buck HJ. The politics of negative emissions technologies and decarbonisation in rural communities. Global Sustainability. 2018;**1**. DOI: 10.1017/sus.2018.2

[9] Ziolo M, Kluza K, Spoz A. Impact of Sustainable Financial and Economic Development on Greenhouse Gas Emission in the Developed and Converging Economies. Energies. 2019; **12**:4514. DOI: 10.3390/en12234514

[10] Chindarkar N, Thampapillai DJ. Rethinking Teaching of Basic Principles of Economics from a Sustainability Perspective. Sustainability. 2018;**10**: 1486

[11] Wagner G. Weitzman, ML. Climate Shock: The Economic Consequences of a Hotter Planet. 2015. Princeton University Press.

[12] Sterner T, Köhlin G. Pricing carbon: Towards a Workable and Effective Climate Regime. CEPR Press and Ferdi. 2019:251-266

[13] Bhandaria V, Giacomonic AM, Wollenberga BF, Wilsona EJ. Interacting policies in power systems: Renewable subsidies and a carbon tax. Electrical Journal. 2017;**30**:80-84

[14] Alton T, Arndt C, Davies R, Hartley F, Makrelov K, Thurlow J, et al. Introducing carbon taxes in South Africa. Applied Energy. 2014;**116**: 344-354

[15] Devarajan S, Go DS, Robinson S, Thierfelder K. Tax policy to reduce carbon emissions in a distorted economy: Illustration from a South African CGE model. The BE Journal of Economic Analysis and Policy. 2011;**11**:1-24

[16] European Union. European Union Trading Systems [internet]. 2018. Available from: https://ec.europa.eu/clima/policies/ets_en [Accessed: 2020/03/19]

[17] Astuti, P, Maryono, M. (2018). Vehicle Emissions Tax: An Opportunity

to Control Air Pollution. E3S Web of Conferences. 73. 10002. 10.1051/e3sconf/20187310002.

[18] Lin B, Li X. The effect of carbon tax on per capita CO2 emissions. Energy Policy. 2015;**39**:5137-5146

[19] Miller, S, Vela, M. Are Environmentally Related Taxes Effective?, Research Department Publications IDB-WP-467; 2013. Inter-American Development Bank, Research Department.

[20] Castiglione C, Infante D, Minervini MT, Smirnova J. Environmental taxation in Europe: What does it depend on? Cogent Economics and Finance. 2014;**2**:1. DOI: 10.1080/23322039.2014.967362

[21] Eurostat. Environmental taxes by economic activity [internet]. 2019. Available from: https://ec.europa.eu/eurostat/cache/metadata/en/env_ac_taxind2_esms.htm [Accessed 2020/04/15].

[22] Voulis, N, van Etten, MJJ, Émile, JL, Chappin, Martijn W, Brazier, FMT. Rethinking European energy taxation to incentivise consumer demand response participation. Energy Policy. 2019;124: 156-168

[23] Delft, CE, Herry, TML., Consult, B. M.E. Case study analysis of the burden of taxation and charges on transport. 2017. Final report, Brussels: European Commission, Directorate-General for Mobility and Transport.

[24] European Commision. Progress made in cutting emissions [internet]. 2020. Available from: https://ec.europa.eu/clima/policies/strategies/progress_en [Accessed 2020/07/20]

[25] European Environment Agency. Total greenhouse gas emission trends and projections in Europe [internet]. 2020. Available from: https://www.eea.europa.eu/data-and-maps/indicators/greenhouse-gas-emission-trends-6/assessment-3 [Accessed: 2020/10/12]

[26] Klier T, Linn J. Using Taxes to Reduce Carbon Dioxide Emissions Rates of New Passenger Vehicles: Evidence from France, Germany, and Sweden. Electron Journal. 2015;7:212-242

[27] Di Cosmo, V, Hyland, M. Carbon Tax Scenarios and their Effects on the Irish Energy Sector. ESRI Working Paper 407, 2011. ESRI, Dublin.

[28] Andersson, JJ. Cars, Carbon Taxes and CO2 Emissions. Centre for Climate Change Economics and Policy Working Paper No. 238, 2017

[29] Bruvoll A, Larsen BM. Greenhouse Gas Emissions in Norway: Do Carbon Taxes Work? Energy Policy. 2015;**32**: 493-505

[30] Revoredo-Giha C, Chalmers N, Akaichi F. Simulating the Impact of Carbon Taxes on Greenhouse Gas Emission and Nutrition in the UK. Sustainability. 2017;**10**:13

[31] González, MB. Implementing a carbon tax in Spain: how to overcome the fear of inflation? Third Annual Conference Fiscal Policies and the Green Economy Transition: Generating Knowledge Creating Impact 29–30 January, 2015 Ca' Foscari University of Venice, Venice, Italy

[32] Haites, E, Maosheng, DG, Kelly, S, Mascher, S, Narassimhan, E and Richards, KR, Wakabayashi, M. Experience with Carbon Taxes and Greenhouse Gas Emissions Trading Systems 2018. Doi.org/10.2139/ssrn.3119241

[33] Ameyaw, B, Yao, L. Analysing the impact of GDP on CO2 emissions and forecasting Africa's total CO2 emissions with non-assumption driven bidirectional long short-term memory. Sustainability. 2018:10. Doi.org/10.3390/su10093110

[34] Asongu S, Montasser GEI, Hassen T. Testing the relationships between energy consumption, CO2 emissions, and economic growth in 24 African countries: a panel ARDL approach. Environmental Science and Pollution Research. 2016;**23**:6563-6573

[35] Aye GC, Edoja PE. Effect of economic growth on CO_2 emission in developing countries: Evidence from a dynamic panel threshold model. Cogent Economics and Finance. 2017;**4**

[36] Farhani S. Rajeb, J B. Link between Economic Growth and Energy Consumption in Over 90 Countries. Interdisciplinary Journal of Contemporary Research in Business. 2012;**3**:282-297

[37] Marx, CK, Slamanig, K. Energy and Carbon Taxes in the EU Empirical Evidence with Focus on the Transport Sector. WIFO Working Papers, No. 555, 2018.

[38] Sterner T. Distributional effects of taxing transport fuel. Energy Policy. 2012;**41**:75-83

[39] Liobikienė G, Butkus M, Matuzevičiūtė K. Environmental Kuznets Curve of greenhouse gas emissions including technological progress and substitution effects. Journal of Energy. 2018;***135***:237-248

[40] Nerudová, D, Hampel, H, Janová, J, Dobranschi, J, Rozmahel, P. 2019. Tax System Sustainability Evaluation: A Model for EU Countries. DOI: 10.1007/s10272-019-0811-6

[41] Park DB, Yoon YS. Developing sustainable rural tourism evaluation indicators. *International* journal of tourism research. 2017;**13**:401-415

[42] Radulescu, M. Sinisi, CI, Popescu, C, Iacob, SE, Popescu, L. Environmental Tax Policy in Romania in the Context of the EU: Double Dividend Theory. Sustainability. 2017: 9:1986.

[43] Streimikiene D, Siksnelyte I, Zavadskas EK, Cavallaro F. The impact of greening tax systems on sustainable energy development in the Baltic States. Sustainability. 2018;**11**:1-29

[44] Kurniawan R, Managi S. Economic growth and sustainable development in Indonesia: an assessment. Bulletin of Indonesian Economic Studies. 2018;**54**: 339-361

[45] Moisescu, E. R. 2015. Analysis of the Relationship Between Sustainable Development and Economic. International Conference on Marketing and Business Development Journal. 2015:1.

[46] Urbaniec, M. Towards Sustainable Development through Eco innovations: Drivers and Barriers in Poland. Economics and Sociology. 2015;8:179–190. DOI:10.14254/2071-789X.2015/8-4/13

[47] Fernando Y, Wah WX. The impact of eco-innovation drivers on environmental performance: Empirical results from the green technology sector in Malaysia. Sustainability production and consumption. 2017;**12**:27-43

[48] Powe N. Sustainable development, sustainability and research within the Journal of Environmental Planning and Management. Journal of Environmental Planning and Management. 2020;**63**: 1523-1527. DOI: 10.1080/09640568.2020.1753409

[49] Aldieri L, Vinci CP. Green economy and sustainable development: the economic impact of innovation on employment. Sustainability. 2018;**10**: 3541. DOI: 10.3390/su10103541

[50] Sauvé, S, Bernard, S, Sloan, P. Environmental sciences, sustainable development and circular economy: Alternative concepts for trans-disciplinary research. Environmental

Development. 2015;17: 48–56. Doi.org/10.1016/j.envdev.2015.09.002.

[51] Kim, MK, Sheu, C, Yoon, J. Environmental Sustainability as a Source of Product Innovation: The Role of Governance Mechanisms in Manufacturing Firms. Sustainability. 2018;10: 1–14. doi.org/10.3390/su10072238

[52] Sarkis, J, Zhu, Q. Environmental sustainability and production: taking the road less travelled. Journal International Journal of Production Research. 2018; 56. Doi.org/10.1080/00207543.2017.1365182

[53] Awan AG. Relationship between environment and sustainable economic development: a theoretical approach to environmental problems. International Journal of Asian Social Science. 2013;**3**: 741-761

[54] Saud AM, Guo P, Haq I, Pan G, Khan A. Do government expenditure and financial development impede environmental degradation in Venezuela? PLoS One. 2019;**1**

[55] Uwazie ONO. Green Economy and Its Implications for Economic Growth in Nigeria. United Nations: Economic Commission for Africa; 2012

[56] You, J, Huang, Y. Green-to-Grey China: Determinants and Forecasts of Its Green Growth. Munich Personal RePEc Archive Paper No. 57468 [internet] 2014. Available from: http://www.mpra.ub.unimuenchen.de/57468 [Accessed 2020/02/22]

[57] Oyebanji IJ, Adeniyi B, Khobai H, Roux LE. Green Growth and Environmental Sustainability in Nigeria. International Journal of Energy Economics and Policy. 2017;7:216-223

[58] Pizer WA, Sexton S. The Distributional Impacts of Energy Taxes. Review of Environmental Economics and Policy. 2017;**13**:104-123

[59] Garidzirai, R, Moyo, C. Tourism, Poverty and Carbon Emissions in Newly Industrialized Countries. Journal of Environmental Management and Tourism. 2020; 11: 653–668. doi.org/10.14505//jemt.11.3(43).19.

[60] Ganda F. The influence of green energy investments on environmental quality in OECD countries. Environmental Quality Management. 2018a;**28**:17-29

[61] Jang EK, Park MS, Roh TW, Han KJ. Policy instruments for eco-innovation in Asian countries. Sustainability. 2015;**7**: 12586-12614

[62] Garidzirai, R. Time Series Analysis of Carbon Dioxide Emission, Population, Carbon Tax and Energy use in South Africa. International Journal of Energy Economics and Policy. 2020; 10: 353–360. Doi.org/10.32479/ijeep.9618

[63] Mohr P. Economics for South African students. Vans Schaik: Pretoria; 2015

[64] Garidzirai R, Chikuruwo RE. An Economic Analysis of the Social Grant Policy in South Africa. Journal of Advanced Research in Law and Economics. 2020, 2020;**11**:362-369

[65] Baltagi, B H. Econometric Analysis of Panel Data. 2nd ed. New York: Wiley, 2001.

[66] Bauma JK, Blanchard W. Inferring shark population trends from generalized linear mixed models of pelagic longline catch and effort data. Fisheries Research. 2003;**102**:229-239

[67] Blundell R, Bond S. GMM Estimation with persistent panel data: an application to production functions. Econometric Reviews. 2000;**19**:321-340

[68] Leve S, Kapingura FM. Financial Development and Income Inequality in the Selected Southern African

Development Community Countries. Journal of Reviews on Global Economics. 2019;**8**:1452-1465

[69] Meraya AM, Dwibedi N, Tan X, Innes K, Mitra S, Sambamoorthi U. The dynamic relationships between economic status and health measures among working-age adults in the United States. Health Economics. 2018;**27**: 1160-1174. DOI: 10.1002/hec.3664

[70] Nayana, S, Kadirb, N, Ahmadc, M, Abdullahd MS. Energy Consumption and GDP: Evidence from Dynamic Panel Data Analysis. International Conference on Economics and Business Research 2013 (ICEBR 2013)

[71] Arellano M, Bond S. Some tests of specification for panel data: Monte Carlo evidence and an application to employment equations. The review of economic studies. 1991;**58**:277-297

[72] Soto M. System GMM estimation with a small sample, No 395, 2009. Working Papers. Barcelona Graduate School of Economics.

[73] Wooldridge, JM. Econometric Analysis of Cross Section and Panel Data. Cambridge, MA: MIT Press; 2002

[74] Wooldridge, J. M. Introductory Econometrics: A Modern Approach. 2nd ed. New York: South-Western; 2003

[75] Ganda F. The effect of carbon performance on corporate financial performance in a growing economy. Social Responsibility Journal. 2018b;**14**: 895-916

[76] Barker T, Junankar S, Pollitt H, Summerton P. Carbon leakage from unilateral environmental tax reforms in Europe, 1995–2005. Energy Policy. 2007;**35**:6281-6292

[77] Acheampong AO. Economic growth, CO2 emissions, and energy consumption: What causes what and where? Energy Economics. 2018;**74**:677-692

[78] Fernández YF, López MF, Blanco BO. Innovation for sustainability: the impact of R&D spending on CO2 emissions. Journal of cleaner production. 2018;**172**:3459-3467

[79] Salahuddin M, Alam K, Ozturk I, Sohag K. The effects of electricity consumption, economic growth, financial development and foreign direct investment on CO2 emissions in Kuwait. Renewable and Sustainable Energy Reviews. 2018;**81**:2002-2010

[80] Guo P, Haq I, Pan G, Khan A. Do government expenditure and financial development impede environmental degradation in Venezuela? PLoS One. 2019;**4**

[81] Zhang YJ, Peng YL, Ma CQ, Shen B. Can environmental innovation facilitate carbon emissions reduction? Evidence from China. Energy Policy. 2017;**100**: 18-28

[82] Ganda F. The impact of industrial practice on carbon emissions in the BRICS: a panel quantile regression analysis. Progress in Industrial Ecology, an International Journal. 2019a;**13**:84-107

[83] Kosonen, K. Regressivity of environmental taxation: myth or reality?. Handbook of Research on Environmental Taxation, Cheltenham: Edward Elgar Publishing, 2012, pp.161–174.

[84] Shafiei S, Salim RA. Non-renewable and renewable energy consumption and CO2 emissions in OECD countries: A comparative analysis. Energy Policy. 2014;**66**:547-556

[85] Blumsack, S, Lave, LB. and Apt, J. Electricity Prices and Costs Under Regulation and Restructuring (August 2008). 2008 Industry Studies Conference Paper, http://dx.doi.org/10.2139/ssrn.1134953

[86] Ganda F. The environmental impacts of financial development in OECD countries: a panel GMM approach. Environmental Science and Pollution Research. 2019b;**26**:6758-6772

[87] Severo EA, de Guimarães JCF, Dorion ECH, Nodari CH. Cleaner production, environmental sustainability and organisational performance: an empirical study in the Brazilian Metal-Mechanic industry. Journal of Cleaner Production. 2015;**96**:118-125

[88] Solaymani S. Carbon and energy taxes in a small and open country. Global Journal of Environmental Science and Management. 2017;**3**(1): 51-62

[89] González R, Hosoda EB. Environmental impact of aircraft emissions and aviation fuel tax in Japan. Journal of Air Transport Management. 2016;**57**:234-240

[90] Magazzino C. The relationship between CO2 emissions, energy consumption and economic growth in Italy. International Journal of Sustainable Energy. 2016;**35**:844-857

[91] Halkos GE, Paizanos EA. The channels of the effect of government expenditure on the environment: evidence using dynamic panel data. Journal of environmental planning and management. 2017;**60**:135-157

[92] Costantini V, Crespi F, Marin G, Paglialunga E. Eco-innovation, sustainable supply chains and environmental performance in European industries. Journal of cleaner production. 2017;**155**:141-154

[93] Phalan B, Ripple WJ, Smith P. Increasing beef production won't reduce emissions. Global change biology. 2016. DOI: 10.1111/gcb.13436

[94] Choi JK, Bakshi BR. Hubacek, K. Nader. A sequential input–output framework to analyse the economic and environmental implications of energy policies: Gas taxes and fuel subsidies. Applied energy. 2016;**184**:830-839

[95] Hatfield-Dodds S, Schandl H, Adams PD, Baynes TM, Brinsmead TS, Bryan BA, et al. Australia is 'free to choose'economic growth and falling environmental pressures. Nature. 2015; **527**:49